Statistical Methods for Field and Laboratory Studies in Behavioral Ecology

Statistical Methods for Field and Laboratory Studies in Behavioral Ecology

Scott A. Pardo
Michael A. Pardo

CRC Press
Taylor & Francis Group
Boca Raton London New York

CRC Press is an imprint of the
Taylor & Francis Group, an **informa** business

A CHAPMAN & HALL BOOK

CRC Press
Taylor & Francis Group
6000 Broken Sound Parkway NW, Suite 300
Boca Raton, FL 33487-2742

First issued in paperback 2020

© 2018 by Taylor & Francis Group, LLC
CRC Press is an imprint of Taylor & Francis Group, an Informa business

No claim to original U.S. Government works

ISBN-13: 978-1-138-74336-6 (hbk)
ISBN-13: 978-0-367-73512-8 (pbk)

Visit the Taylor & Francis Web site at
http://www.taylorandfrancis.com

and the CRC Press Web site at
http://www.crcpress.com

Contents

Preface

Behavioral ecology is a field that largely depends upon empirical investigation and observation, and as such leans heavily on statistical methods. Thus, behavioral ecologists require some instruction into those statistical concepts and methods that will be of use to their work. For example, in the September 2017 issue of the journal *Behavioral Ecology*, we counted more than 50 different statistical techniques. Although all the methods mentioned in that work could be found in various statistical texts, it would be difficult to find all of them in one place. This text was written with the behavioral ecologist in mind. Not only does it contain information on methods that have been widely used by behavioral ecologists, it also provides a little depth into the theory under which those methods were derived. Without getting overly mathematical, the theoretical aspects are described in order to elucidate the assumptions and limitations of the methods. In that way, the scientist will have a better view as to when these methods are applicable, and the appropriate level of skepticism required when interpreting results. Sometimes there may be more than one technique for analyzing the same data and providing the same type of conclusion. This text will also compare such methods, describe their assumptions, and hopefully provide some insight into which technique the researcher might choose. In particular, methods that require few assumptions about the underlying probability distributions of populations or data-generating processes will be described, together with associated computer programs. The computer programs provided are written in the R language, which has gained much popularity in the scientific world. Datasets provided are mostly based, at least to some degree, on real studies, but the data themselves are simulated, and the examples are simplified for pedagogical purposes. Those studies providing the inspiration for the simulated data are cited in the text.

It is assumed that the reader has had exposure to statistics through a first introductory course at least, and also has sufficient knowledge of R. This is not a primer for R or for statistics. However, some introductory material is included to aid the less initiated reader. The first five chapters largely consist of material covered in many first courses on statistics for biologists. However, there is mention of some intermediate notions, such as rank-based methods, permutation tests, and bootstrapping. In most chapters, at least two different methods are presented, together with their primary assumptions, for analyzing the exact same data. As such, this is not a book about parametric, nonparametric, frequentist, or Bayesian statistics. Rather, with no sword to grind, statistical methods are presented to the researcher in order to familiarize him or her with techniques described in scholarly literature.

Hopefully, the text will remove the perception of the magical aura that statistical methods often evoke.

The remaining chapters cover methods that each have multiple books written on them. As such, this can only be viewed as an introduction, and an introduction to some more fundamental but not elementary methods. Nevertheless, the material presented should at least get the reader started on the path.

Something should be said about the organization of material within a chapter. Except for Chapters 1, 15, and 16, each chapter is divided into five sections:

General Ideas

Examples with R Code

Theoretical Aspects

Key Points

Exercises and Questions

Hopefully, the first two sections, General Ideas and Examples with R Code, can get the reader started in the process of analyzing data. The Theoretical Aspects section will help provide some explanation of how the methods actually work, why they work, and what assumptions are necessary for them to work correctly. We strongly recommend that the student reads the Theoretical Aspects sections in order to gain a better understanding of the methods, their strengths, and their limitations.

As in the case of all texts, some very important topics have been omitted. In particular, the uses of statistical methods for phylogenetic analyses and spatial modeling have not been discussed. These, and other advanced methods, are beyond the scope of this book.

Acknowledgments

The authors would like to acknowledge and thank Yehudah A. and Jeremy D. Pardo for their unwitting contributions to this work. The authors would also like to thank Maria Modanu and Sarah Bluher for their review, explanations, and suggestions about decision and game theory. Scott would like to acknowledge and thank Dr. Rezi Zawadzki for her encouragement and suggestions. Finally, both authors, Scott and Michael, owe a great debt to their wife and mother, respectively, for first of all giving them the idea and suggesting they write this book, and for continually making suggestions during the writing. They especially owe her for conceiving of and suggesting the last chapter, which would not exist if she hadn't thought of it, and if she hadn't persisted in encouraging its writing.

About the Authors

Scott Pardo has been a professional statistician for 37 years. He has worked in a diverse set of environments and applications, including the U.S. Army, satellite communications, cardiac pacemakers, pharmaceuticals, and blood glucose meters. He has a PhD in systems and industrial engineering from the University of Southern California, is a Six Sigma Master Blackbelt, and an accredited professional statistician, PStat®.

Michael Pardo is a PhD candidate in behavioral ecology at Cornell University, and has been conducting field-based research in animal behavior for over 10 years. He holds a BS in environmental biology from the State University of New York (SUNY) College of Environmental Science and Forestry. His primary research interests are in vocal communication and social cognition, particularly with mammals and birds. He has studied eastern gray squirrels, Asian elephants, and acorn woodpeckers.

1

Statistical Foundations

Statistics has its foundation in probability. The basic building block is known as the random variable. Without being overly mathematical, random variables are those things that can be expressed in some sort of quantitative fashion, and whose values cannot be perfectly predicted. Random variables will take the form of observations or measurements made on experimental units. Experimental units are very often individual animals, but could be a collective, such as a flock, herd, hive, family, or other collection of individuals. The observations and measurements to be discussed in this text will be things that can be quantified. For example, a variable might have only two possible values: Say, one, if a particular, predefined behavior is observed under particular conditions; and zero if it is not. Another variable could be the distance traveled by an individual in some fixed period of time. The random nature of these variables implies that they have a probability distribution associated with their respective values. The analyses of data will be all about features of these distributions, such as means, standard deviations, and percentiles.

By way of a taxonomy for observations or measurements, we will refer to those whose values can be expressed as an integer as *discrete,* and those whose values can be expressed as a decimal number or fraction as *continuous.* Analyses for these types of variables are different in details, but have similar aims.

Statistical analyses involve three basic procedures:

1. Estimation
2. Inference and decision making
3. Model building: Discrimination and prediction

In all cases, statistics is the science of applying the laws and rules of probability to samples, which are collections of values of a random variable or in fact a collection of random variables. The type of sample upon which we will most heavily rely is called the *random sample.* A random sample can be defined as a subset of individual values of a random variable where the individuals selected for the subset all had an equal opportunity for selection. This does not mean that in any given data-gathering exercise there could not be more than one group or class of individuals, but that within a class the individuals chosen should not have been chosen with any particular bias.

The nature of all three types of procedures can be subdivided into two basic classes:

1. Parametric
2. Nonparametric

By parametric, we mean that there is some underlying "model" that describes the data-generating process (e.g., the normal, or Gaussian, distribution), and that model can be described by a few (usually one to three) numerical parameters. By nonparametric, we mean analyses that are not dependent on specifying a particular form of model for the data-generating process. Both paradigms for statistical analyses are useful and have a place in the data analyst's toolbox. As such, both classes of analyses will be discussed throughout the text.

Some Probability Concepts

Parametric distributions are described by mathematical functions. The fundamental function is called the *probability density function for continuous variables,* or in the case of discrete variables, it is often called the *probability mass function.* The idea is to describe the probability that the random variable, call it X, could take on a particular value, or have values falling within some specified range. In the case of continuous variables, the probability that X is exactly equal to a particular value is always zero. This rather curious fact is based on a set of mathematical ideas called *measure theory.* Intuitively, the probability of finding an individual with exactly some specific characteristic (say, a weight of 2.073192648 kg) is, well, zero. This is not to say that once you find such an individual, you must be hallucinating. The notion of zero probability (and in fact any probability) relates to a priori determination, that is, before any observation. Once an observation is made, the probability of observing whatever it is you observed is in fact 1, or 100 percent.

In general, capital letters, like X, will refer to a random variable, whereas lower case letters, like x, will refer to a specific value of the random variable, X. Often, in order to avoid confusing discrete and continuous variables, the symbol $f_X(x)$ will refer to the density function for variable X, evaluated at the value x, and $p_X(x_k)$ to a probability mass function for a discrete variable X evaluated at the value x_k. The notation $Pr\{\}$ will refer to the probability that whatever is inside the curly brackets will happen, or be observed. If the symbol "dx" means a very small range of values for X, and x_k represents a particular value of a discrete random variable, then

$$f_X(x)dx = Pr\{x - dx \leq X \leq x + dx\}$$

and

$$p_X(x_k) = \Pr\{X = x_k\}$$

There is a particularly important function called the *cumulative distribution function* (CDF) that is the probability $\Pr\{X \leq x\}$, which is usually defined in terms of density or mass functions, namely

$$F_X(x) = \int_{-\infty}^{x} f_X(\xi)d\xi$$

for continuous variables, and

$$F_X(x) = \sum_{x_k \leq x} p_X(x_k)$$

for discrete variables.

As mentioned earlier, the functions $f_X(.)$ and $p_X(.)$ generally have parameters, or constants, that dictate something about the particular nature of the shape of the density curve. Table 1.1 shows the parameter lists, density or mass functions for several common distributions.

In the case of the binomial and beta distributions, the symbol p was used to denote a parameter (binomial), or as a value of a random variable (beta), and not the mass function itself. The function $\Gamma(x)$ is called the gamma function (oddly enough) and has a definition in terms of an integral:

$$\Gamma(x) = \int_{0}^{\infty} \xi^{x-1}e^{-x}d\xi$$

Aside from the CDF, there are some other important functions of $f_X(x)$ and $p_X(x_k)$. In particular, there is the expected value, or mean:

$$E[X] = \mu = \begin{cases} \sum_k x_k p_X(x_k) \\ \int_{-\infty}^{+\infty} \xi f_X(\xi)d\xi \end{cases}$$

and the variance:

TABLE 1.1

Some Probability Density and Mass Functions

Name	Parameters	Density or Mass Function	Range of Values
Normal	μ, σ	$\dfrac{1}{\sqrt{2\pi}\sigma}\exp\left(-\dfrac{1}{2}\left(\dfrac{x-\mu}{\sigma}\right)^2\right)$	$-\infty < x < +\infty$
Gamma	n, λ	$\dfrac{\lambda^n}{\Gamma(n)}x^{n-1}\exp(-\lambda x)$	$x > 0$
Chi-Squared	ν	$\dfrac{(1/2)^{\frac{\nu}{2}}}{\Gamma\left(\dfrac{\nu}{2}\right)}x^{\frac{\nu}{2}-1}\exp\left(-\dfrac{1}{2}x\right)$	$x > 0$
Student's t	ν	$\dfrac{\Gamma\left(\dfrac{1}{2}(\nu+1)\right)}{\sqrt{\pi\nu}\,\Gamma\left(\dfrac{1}{2}\nu\right)}\left[1+\dfrac{x^2}{\nu}\right]^{-\frac{(\nu+1)}{2}}$	$-\infty < x < +\infty$
F	ν_1, ν_2	$\dfrac{\Gamma\left(\dfrac{\nu_1+\nu_2}{2}\right)}{\Gamma\left(\dfrac{1}{2}\nu_1\right)\Gamma\left(\dfrac{1}{2}\nu_2\right)}\dfrac{x^{\left(\frac{\nu}{2}\right)-1}}{(1+x)^{(\nu_1+\nu_2)/2}}$	$x > 0$
Poisson	λ	$\dfrac{\lambda^k e^{-\lambda}}{k!}$	$k = 0, 1, 2, \ldots$
Binomial	n, p	$\binom{n}{k}p^k(1-p)^{n-k}$	$k = 0, 1, 2, 3,\ldots n$
Beta	α, β	$\dfrac{\Gamma(\alpha+\beta)}{\Gamma(\alpha)\Gamma(\beta)}p^{\alpha-1}(1-p)^{\beta-1}$	$0 < p \le 1$

$$V[X] = E[X-\mu^2] = \sigma^2 = \begin{cases} \displaystyle\sum_k (x_k-\mu)^2 p_X(x_k) \\ \displaystyle\int_{-\infty}^{+\infty} (\xi-\mu)^2 f_X(\xi)d\xi \end{cases}$$

Commonly the Greek letter μ is used to symbolize the expected value, and σ^2 is used to represent the variance. The variance is never negative (it is a sum of squared values). The square root of the variance is called the standard deviation, and has its most important role in random variables having a normal distribution. The expected value has units that are the same as individual measurements or observations. The variance has squared units, so that the standard deviation has the same units as the measurements.

Often we must deal with more than one random variable simultaneously. The density or mass function of one variable might depend on the value of some other variable. Such dependency is referred to as *conditioning*. We symbolize the conditional density of X, given another variable, say Y, is equal to a particular value, say y, using the notation:

$$f_{X|Y}\left(x|Y=y\right)$$

Typically, the fact that $Y = y$ will affect the particular values of parameters. Also, we will usually drop the subscript $X|Y$, since the conditional nature of the density is made obvious by the "|" notation.

It is possible that the value of one random variable, say Y, has no effect on the probability distribution of another, X. It turns out that any two random variables have what is called a joint density function. The joint density of X and Y could be defined as

$$f_{XY}(x,y)dxdy = Pr\{x-dx \le X \le x+dx, \; AND \; y-dy \le Y \le y+dy\}$$

The joint density quantifies the probability that random variable X falls in a given range and at the same time random variable Y falls in some other given range.

It turns out that this joint density can be expressed in terms of conditional densities:

$$f_{XY}(x,y) = f_{X|Y}(x|y)f_Y(y)$$

The marginal density of one variable (say X) is the density of X without the effect of Y, and is computed as

$$f_X(x) = \int_{-\infty}^{+\infty} f_{XY}(x,y)dy$$

When X and Y are independent of each other, then

$$f_{X|Y}\left(x|y\right) = f_X(x)$$

So that

$$f_{XY}(x,y) = f_X(x)f_Y(y)$$

In other words, when X and Y are independent, their joint density is the product of their marginal densities.

In addition to joint distributions, the expected values and variances of sums and differences of random variables find themselves in many applications. So, if X and Y are two random variables:

$$E[X \pm Y] = E[X] \pm E[Y]$$

If X and Y are independent, then

$$V[X \pm Y] = V[X] + V[Y]$$

While the sign of the operator ($\pm$) follows along with the expected values, the variance of the difference is the sum of the variances.

Another set of facts we will use relating to conditional densities or mass functions is based on something called Bayes' theorem. Briefly, Bayes' theorem states that if X is a random variable with density f, and Y is a random variable with density g, then

$$g(x|Y = y) = \frac{g(y)f(x|Y = y)}{\int_{-\infty}^{+\infty} f(x|Y = \xi)g(\xi)d\xi}$$

As long as Y is continuous, this particular formula holds even if X is discrete, and f is the mass function of X. If, however, Y is discrete, and g is its mass function, then the integral is replaced with a summation:

$$g(x|Y = y) = \frac{g(y)f(x|Y = y)}{\sum_{k} f(x|Y = \xi_k)g(\xi_k)}$$

It should be noted that it is possible for a random variable to not actually have a density function associated with it. However, that situation probably never exists in nature, so we will assume the density always exists.

Some Statistical Concepts

Earlier we mentioned that statistical problems could be classified into the categories:

1. Estimation
2. Inference and decision making
3. Model building: Discrimination and prediction

Estimation is the process of using data to guess at the value of parameters or some feature of a probability distribution assumed to be governing the data-generating process. Probably the most common is estimating the expected value of a distribution. The expected value of the random variable's distribution is

$$E[X] = \mu = \begin{cases} \sum_k x_k p_X(x_k) \\ \int_{-\infty}^{+\infty} \xi f_X(\xi) d\xi \end{cases}$$

One of the useful mathematical properties of expected value is that it is a linear operator, namely:

$$E[X_1 + X_2 + \cdots + X_n] = E[X_1] + E[X_2] + \cdots + E[X_n]$$

and

$$E[aX] = aE[X]$$

when a is a non-random constant. An estimate based on a sample of observations from the data-generating process is

$$\hat{\mu} = \frac{1}{n} \sum_{i=1}^{n} x_i$$

We use the notation $\hat{\mu}$ instead of a perhaps more well-recognized symbol $\bar{x}$, to emphasize the fact that we are using the data to estimate the expected value. There are many such estimation formulae (called estimators), and many are used in different contexts for different reasons. The main point is that data can be used to estimate parameters or other features of probability distributions. The other point is that, since estimators use data, they themselves are random variables. Thus, if two researchers studying the same population of finches each make independent observations on either two sets (samples) of birds or even on the same sample, but at two different times, and each researcher calculates an average, the two averages most likely won't be exactly the same.

There are different methods used to derive estimator formulas for various parameters. Perhaps the best known is called the method of maximum likelihood. The idea is that if you have a random sample of measurements (X), you can find values of parameters that maximize something called the like-lihood function, which generally depends on assuming the form of the dis-tribution for the data-generating process. Suppose that the values x_1, x_2, ..., x_n represent n values sampled from a normally distributed data-generating process, with unknown expected value and variance the density function evaluated at x_i, say, would be given by

$$f(x_i) = \frac{1}{\sigma\sqrt{2\pi}} e^{-\frac{1}{2}\left(\frac{x_i-\mu}{\sigma}\right)^2}$$

The likelihood function for the sample would be the product of all the valuations of the density function:

$$L(x_1, x_n, ..., x_n) = \prod_{i=1}^{n} f(x_i)$$

Of course, this likelihood function cannot be computed without know-ing μ and σ. The idea of maximum likelihood is to find values $\hat{\mu}$ and $\hat{\sigma}$ that maximize L. Usually the log of the likelihood function is taken before attempting to solve the maximization. Maximizing the log of L is equivalent to maximizing L, since the log is a monotonic increasing function. The log of a product is the sum of the logs of the factors:

$$\log L = \sum_{i=1}^{n} \log(f(x_i))$$

Maximizing the sum is easier mathematically than maximizing the product.

What is important to note is that first we had to pick a parametric form for the density function of the random variable from which we were sampling, the parameter values are unknown, and our guess for the parameter values is based on a criterion that gives us the best guess. It turns out that for the normal model, the maximum likelihood estimators for μ and σ^2 are

$$\hat{\mu} = \frac{1}{n}\sum_{i=1}^{n} x_i$$

and

$$\hat{\sigma}^2 = \frac{1}{n}\sum_{i=1}^{n}\left(x_i - \hat{\mu}\right)^2$$

Some may notice that the maximum likelihood estimator for σ^2 differs from the formula used in most elementary texts, in that it divides by n and not $n - 1$. Dividing the sum by $n - 1$ to estimate σ^2 gives the formula a property known as *unbiasedness*. While this is important, in the case of this estimator the effects are fairly small. Another estimation method is called *least squares*. Rather than maximize a likelihood function, least squares chooses estima-tors that minimize an "error" function. A common context for least squares estimation is linear regression. More will be said about least squares. For now, just recognize it as a method for estimating parameters.

Statistical estimates, since they are based on a finite sample of observa-tions or measurements made on individuals taken from some population or data-generating process, have themselves a random variation component. Inasmuch as a statistical estimate is attempting to help make a guess about a parameter, it would be good to know that the formula used to compute the estimate has a reasonable chance of getting close to the actual value of the parameter. One such property has already been described, namely, maxi-mum likelihood. Another property that is desirable is unbiasedness, which was also mentioned earlier. An estimation formula is said to be unbiased if its expected value is equal to the parameter to be estimated. For example, assuming a random sample, $x_1, x_2, \ldots, x_n$, then the expected value of each x_i is the population mean, μ, and

$$E[\hat{\mu}] = \frac{1}{n}\sum_{i=1}^{n}E[x_i] = \frac{1}{n}\sum_{i=1}^{n}\mu = \mu$$

Thus, our arithmetic mean estimator for μ is in fact unbiased. Conversely, the maximum likelihood estimate of σ^2 is not unbiased (or, in other words, biased). It turns out that

$$E[\hat{\sigma}^2] = \frac{1}{n}\sum_{i=1}^{n}E\left[(x_i - \hat{\mu})^2\right] = \frac{n-1}{n}\sigma^2$$

Thus, the maximum likelihood estimator of σ^2 slightly underestimates the variance. The point of the discussion about unbiasedness is that estimation formulae are themselves random variables, and as such we will need to con-sider their probabilistic characteristics.

Inference is about making a priori guesses about parameter values and then using data to decide if you were correct. Suppose, for example, you guessed that the average duration of a courtship display was 30 seconds. How would you decide whether to believe your guess, or not? First you would gather data, by timing courtship displays of several individuals, say n. Then you would probably compute the maximum likelihood estimates of mean and variance. Suppose the estimate of the mean was 31.5 seconds, and the standard deviation (square root of variance) estimate was three seconds. OK, so it wasn't 30. Were you wrong? The question becomes one of how much variation there might be if the experiment were repeated. The idea of statistical inference is to make a decision about what to believe, and not what actually is the truth. Our decision has risk associated with it, namely the risk (or probability) of saying our guess is wrong when in fact it is correct, and the risk of saying our guess is correct when in fact it is not. There is a formalism for expressing the notions of inference. There are two competing hypotheses, or guesses, about the parameter or parameters of interest. One is called the "null" hypothesis, symbolized as H_0. The logical negation of the null hypothesis is called, not surprisingly, the alternate hypothesis, and is often symbolized as H_1. So, in the example of the courting display question, we might have

$$H_0: \mu = 30$$
$$H_1: \mu \neq 30$$

The error of deciding that H_0 is false when in fact it was true is called a Type I error. The error of believing H_0 is true when it is not is called a Type II error. The next thing required is a rule, based on data, that lets the decisionmaker decide whether to believe H_0 or H_1. Since data are subject to variation, the rule is necessarily probabilistic. It turns out that, conveniently, the calculation

$$t = \frac{\hat{\mu} - 30}{\hat{\sigma}/\sqrt{n}}$$

has a known probability distribution, the familiar Student's t, provided that the null hypothesis is actually correct (i.e., that $\mu = 30$). This formula is known as a *test statistic*, because it is the quantity we will use to decide whether to believe (accept) the null hypothesis, or disbelieve it (reject). In fact, a common feature of all inference is determining the distribution of the test statistic if H_0 were actually true. The probability of making a Type I error is symbolized with the letter a. The probability of a Type II error is traditionally symbolized with the letter β. We can find a range of values that t would fall in between with probability $1 - \alpha$, given that H_0 is true, even before we gathered any data. In fact, the range of possible values only depends on the sample size, n, and the desired probability content of the range. If, for example, the sample size was $n = 10$, and we wanted the probability content to be $100(1 - \alpha)\% = 95$ percent, then the range of values for t we would expect if the null hypothesis was correct would be approximately ± 2.228. The range ($t \leq$ -2.228, $t \geq +2.228$) is called the *critical region* of size α. If the value of the test statistic falls in the critical region, we say that the test statistics is significant, and we REJECT the null hypothesis is favor of the alternative. The particular region for this example is partially based on the presumption that we computed the maximum likelihood estimate for standard deviation. If after getting data we computed the value of t using the formula above, and its value fell within the range ± 2.228, we would continue to believe the null hypothesis, because there is a fairly "high" (95 percent) chance of t falling inside this range if H_0 is correct. Conversely, there is a relatively "low" chance that t would fall outside the "critical" range if H_0 was correct. Unfortunately, we cannot make the same statement about the alternative hypothesis, H_1, since there are an infinite number of possible values (anything other than 30) that would make it correct. Thus, it is easier to fix the chance of making the mistake of deciding that H_0 is false when in fact it is true. Once this risk is decided upon, the decision rule for either believing H_0 or not believing it is fairly easy to compute, provided we know something about the distribution of the test statistic, given that the null hypothesis is true.

Another way of determining a rule for rejecting or accepting the null hypothesis is to *compute a probability* of observing the data you got IF the null hypothesis was actually correct. This probability is usually referred to as a *p-value*. Thus, in our example, if in fact $\mu = 30$, then the test statistic

$$t = \frac{\hat{\mu} - 30}{\hat{\sigma}/\sqrt{n}}$$

has a Student's t distribution with degrees-of-freedom parameter equal to n (since we used the maximum likelihood estimate of σ). Suppose we had data that yielded a sample estimate of μ, say,

$$\hat{\mu} = \frac{1}{n}\sum_{i=1}^{n} x_i = 31.5$$

and an estimate of σ^2:

$$\hat{\sigma}^2 = \frac{1}{n}\sum_{i=1}^{n}(x_i - \hat{\mu})^2 = 9$$

If $n = 25$, then the sample test statistic would be

$$t = \frac{31.5 - 30}{3\Big/\sqrt{25}} \approx 2.50$$

Since the alternative hypothesis is $\mu \neq 30$, we compute the probability that the test statistic would be outside the range $(-2.50, +2.50)$. To compute this, we can use the R function $pt()$:

$$\text{pt}(q = -2.5,\ df = 25,\ \text{lower.trail} = \text{TRUE}) \approx 0.01934$$

and

$$\text{pt}(q = +2.5,\ df = 25,\ \text{lower.trail} = \text{FALSE}) \approx 0.01934$$

The "two-sided" p-value is $0.01934 + 0.01934 = 0.03868$.

Since Student's t distribution is symmetric about zero, the probability for the "lower tail" of $-t$ is equal to the probability for the "upper tail" of $+t$.

If our threshold of p-values is $\alpha = 0.05$, then since 0.03868 is less than 0.05, we will no longer believe that the null hypothesis is correct, and reject it. With a sample size of $n = 25$, and $1 - \alpha = 0.95$, then the critical region is $t \leq -2.06$, $t \geq +2.06$. Since $t = 2.50 > +2.06$, we would reject the null hypothesis. Regardless of whether you determine a critical region of size α, or choose α to be a threshold for p-values, the conclusions would be identical.

Another methodology that is somewhere between estimation and inference is called confidence interval building. The confidence interval again employs that risk level, α, but in a slightly different manner. Suppose we wanted to know that value of the parameters that would correspond to the limits of the critical range for the test statistic. Using the previous example, let

$$t_{low} = -2.06 = \frac{\hat{\mu} - \mu_{low}}{\hat{\sigma}\Big/\sqrt{n}}$$

and

$$t_{high} = -2.06 = \frac{\hat{\mu} - \mu_{high}}{\hat{\sigma}\Big/\sqrt{n}}$$

Solving for μ_{low} and μ_{high} gives:

$$\mu_{low} = \hat{\mu} - 2.06\frac{\hat{\sigma}}{\sqrt{n}}$$

and

$$\mu_{high} = \hat{\mu} + 2.06\frac{\hat{\sigma}}{\sqrt{n}}$$

The range of values (μ_{low}, μ_{high}) is called the $100(1 - \alpha)\%$ "confidence interval" for parameter μ. It can be thought of as a feasible range for the unknown values of μ. That is, we are not certain about the actual value of μ, but we are nearly certain ($100(1 - \alpha)\%$ certain) that it lies somewhere in the interval (μ_{low}, μ_{high}). So, in our example with $\hat{\mu} = 31.5$, $\hat{\sigma} = 3$, and $n = 25$, the 95-percent confidence interval would be

$$\mu_{low} = 31.5 - 2.06\frac{3}{\sqrt{25}} \approx 30.26$$

$$\mu_{high} = 31.5 + 2.06\frac{3}{\sqrt{25}} \approx 32.74$$

Since the hypothetical value for μ, namely 30, is not contained in the confidence interval (30.26, 32.74), we do not believe that 30 is a feasible value for μ.

When the null hypothesis is rejected, we say that the difference between our estimate of the parameter and the null value is *statistically significant at the $100\alpha\%$ level*. Another way of stating the same thing is that if we reject the null hypothesis, we would believe that the results of our analyses are repeatable.

Model building is a special application of estimation, but it usually has some inference associated with it. The idea is to postulate some mathematical relationship between some variables, some random and some without any random component. Then we estimate the values of model parameters. Finally, we test to see if we should believe that the form of the model we postulated was reasonable. Models can be predictive or discriminatory/classificatory. A simple example of a predictive model would be a simple linear regression. Suppose there is a continuously valued random variable, Y, and another continuously valued nonrandom variable, X. Y could be things such as response time, elapsed time, distance traveled, or other random variables that can be expressed as a decimal number. In this simple case, we are assuming the X variable is not random. In other words, X is something whose value would have no random variation, and whose value is known perfectly without error. Y is referred to as the response variable, and X is the predictor or regressor variable. The general form of the linear model is

$$Y = \beta_0 + \beta_1 X + \varepsilon$$

The coefficients β_0 and β_1 are unknown, and need to be estimated. The variable ε represents random "noise," indicating that the value of Y is on the average a linear function of X, but the actual observed values may have some perturbations, or noise, or sometimes-called errors associated with them.

Once the values of the parameters are estimated, a predicted value of Y can be computed for a given value of X. We would not usually consider X to have random noise associated with it. That is, when we get a value for X, we are (mostly) certain that the value would not vary if we measured or observed it a second time under exactly the same conditions. Rather, we suppose that given the value of X, we can predict on the average what Y would be, with the understanding that Y might vary from this average.

Another closely related type of model is also linear, but is classificatory or discriminatory. The X variables are not continuous, but are discrete categories. The goal is to determine if particular groupings of individuals actually discriminate between individuals. In other words, we want to know if individuals in different groups actually differ from each other with respect to Y. Perhaps the simplest example is the one-way analysis of variance (ANOVA). In this case, the single X variable is a set of discrete categories, and Y is the continuous random variable response. The question is not to find a prediction of Y for a given value of X, per se. Rather, the question is to estimate the difference in the average Y between the different categories. In the case of ANOVA, often the inferential part of modeling is of greater interest, namely, whether the difference in average values of Y between the different groups of X categories is in fact repeatable.

There are certainly more types of both predictive and classificatory modeling. The key notion here is that data can be used to create these sorts of models, through a combination of estimation and inference.

This is the classical parametric methodology for statistical inference. There is another set of methods, sometimes called nonparametric or distribution-free, of which neither term is strictly true. The idea is that the distribution of test statistics should not depend on the distribution of the data-generating process. The basic idea is still the same; you formulate a test statistic, you determine the "critical range" or "critical value" based on α, you get some data, and then you compute the test statistic to decide if you should accept or reject the null hypothesis.

A special set of nonparametric techniques is sometimes referred to as *resampling methods*. This book will in fact emphasize resampling methods where appropriate. The resampling techniques will generally fall into the bootstrap estimation process or the permutation hypothesis testing process. Both of these methods are computer-based, but given modern computing software such as R, they are fairly easy to perform.

Bayesian statistics is an alternate view of parameters, not as particular values to estimate or about which to make a guess about their true values, but treating them as if they themselves are random variables. Like the classic "frequentist" approach, Bayesian methods employ a likelihood function. However, these methods incorporate prior information about the parameters of interest. "Prior" to making observations, the analyst posits a distribution of the parameters of interest. The "prior" distribution expresses the knowledge about the parameter prior to performing the "next" experiment. So, for example, perhaps the mean response time to a stimulus is guessed to be most likely 10 seconds, but it could be as fast as 5 seconds and as delayed as 15 seconds. Rather than simply hypothesizing that the mean is exactly 10 seconds, the Bayesian method is to postulate a distribution that expresses the current level of knowledge and uncertainty in the parameter. Then, once data are gathered, Bayes' theorem is used to combine the prior distribution with the likelihood function, to update the prior knowledge. The updated distribution for the parameter is called the posterior distribution. So, if $f_{old}(\tilde{\mu})$ represents the prior density function for the parameter $\tilde{\mu}$, and $L[x_1, x_2, ..., x_n | \tilde{\mu}]$ the likelihood function for the sample, given a particular value of $\tilde{\mu}$, then the updated density function (called the posterior density) is

$$f_{new}(\tilde{\mu} | x_1, x_2, ..., x_n) = \frac{f_{old}(\tilde{\mu})L[x_1, x_2, ..., x_n | \tilde{\mu}]}{\int_{-\infty}^{+\infty} f_{old}(\xi)L[x_1, x_2, ..., x_n | \xi]d\xi}$$

Key Points for Chapter 1

- The primary concept for probability in our context is the random variable; it is generally defined in terms of measurements or observations made, where the values of those measurements or observations cannot be deduced exactly before they are made.

- Random variables come in two flavors: discrete and continuous.

- While the actual value of a random variable cannot be known a priori, statements can be made about the probability that a random variable will have a certain value, or its value will fall within some specific range.

- The density and probability mass functions are the means by which all probabilities about random variables are calculated; most (in our case, all) of these functions have a small number of parameters (one or two usually) whose values define the exact shape of the function.

- The densities are used to derive the average, or expected, value of the random variable, and the variance, a measure of spread for the density.

- Two random variables will have a joint density, and each will have a conditional and marginal density.

- Bayes' theorem is a means of finding a conditional density of one variable given the value of another.

- Data are used to estimate the values of unknown parameters; a classic estimation approach is to find the parameter value that maximizes the likelihood function, which is a sort of joint density function.

- Inference is about deciding based on data whether a guess about the value of a parameter is believable.

Exercises and Questions

1. Explain the differences in the nature of probability statements you can make concerning discrete versus continuous random variables.

2. In a sample of 10 northern shovelers, the mean bill length was 6.25 centimeters, with a standard deviation of 1.10 centimeters. Test the hypothesis that the expected value of bill length is 6.00 centimeters; use $\alpha = 0.05$, with a two-sided test. Compute a two-sided 95-percent confidence interval for mean bill length.

3. The percent of male chickens (R for roosters) in a farm population is 10 percent; the females (H for hens) make up 90 percent. The probability of a certain aggressive behavior (A) for males is 80 percent; for females, the probability is 10 percent. What is the probability of observing the aggressive behavior, regardless of the sex of the individual chicken?

2

Binary Results: Single Samples and 2 × 2 Tables

General Ideas

Single Proportion

A random variable is called binary if it has only two possible values, usually assigned the numbers zero or one. Typically, the variable has a definition like

$$X = \begin{cases} 1 & \text{if observation has characteristic of interest} \\ 0 & \text{otherwise} \end{cases}$$

Assuming that the probability of observing the characteristic in an individual experimental unit is constant (at least over some time and space region), then X has something called a *Bernoulli distribution*, which is defined with a single parameter, p, as

$$Pr\{X = 1\} = p; \ Pr\{X = 0\} = 1 - p$$

If a random sample of n Bernoulli variables is obtained and a new variable is defined to be the number of values equal to one out of the n observations, then this new variable, call it Y, has a binomial distribution. The general idea is that the data are a sample of results each having only one of two possible states or values (e.g., individual observed to exhibit a particular behavior or not). The methods are designed to test hypotheses about the parameter p, the probability that the variable would have one of the two possible states in a randomly selected individual. Of course, the researcher may simply want to compute a confidence interval for the parameter p. We will focus on computing a confidence interval.

A simple way of computing a confidence interval for p is based on the formulas derived by Clopper and Pearson (1934). The formulas are

$$p_L = \frac{(2y+2)F_{\alpha/2}^{-1}(2y+2,\ 2n-2y)}{2(n-y)+2(y+1)F_{\alpha/2}^{-1}(2y+2,\ 2n-2y)}$$

and

$$p_U = \frac{(2y+2)F_{1-\alpha/2}^{-1}(2y+2,\ 2n-2y)}{2(n-y)+2(y+1)F_{1-\alpha/2}^{-1}(2y+2,\ 2n-2y)}$$

That is, the confidence interval would be $(p_L,\ p_U)$. The notation F_r^{-1} (*num,denom*) refers to the value (percentile) of the F distribution corresponding to probability r and having numerator degrees of freedom *num* and denominator degrees of freedom *denom*.

2 × 2 Tables

Suppose that once again the measure of goodness is Bernoulli, but there are either two groups or two conditions we would like to compare. Let X = 1 if the characteristic of interest is observed, and X = 0 otherwise. The objective is to decide whether $Pr\{X = 1\}$ is different between the two groups or conditions (which we will often call treatments, even if the condition is not about treating anything). Furthermore, suppose there are two independent samples, one for group #1 and one for group #2. Then the results of the observations could be summarized in a table with two rows and two columns (hence the 2 × 2 designation). Such tables are referred to as 2 × 2 contingency tables. Table 2.1 shows a sort of generic 2 × 2 contingency table.

If the proportion of ones for Group #1 is X1/N1, and X2/N2 for Group #2, the question is whether this pattern is repeatable or not. One way to determine repeatability is to get a *p*-value for the table, under the null hypothesis that there is no difference in the probability that X would equal one for Group #1 or Group #2. A popular test is called Fisher's exact test, since it relies on the exact distribution of X1 and X2, assuming the probability that X = 1 is constant.

TABLE 2.1

A Generic 2 × 2 Table

	X = 1	X = 0	Row Total
Group #1	X1	N1 – X1	N1
Group #2	X2	N2 – X2	N2
Column Total	X1 + X2	N1 + N2 – (X1 + X2)	N1 + N2

Examples with R Code

Single Proportion

Figure 2.1 shows R code for computing of confidence intervals for proportions. Table 2.2 shows the output for two examples, one with $Y = 17$ characteristics of interest observed out of $n = 20$ and another with $Y = 85$, $n = 100$.

As a more concrete example, suppose that out of $n = 100$ individual New Caledonian crows (*Corvus moneduloides*), 93 were observed to use a stick to obtain food. The Clopper–Pearson 95-percent confidence interval for p, the probability that an individual New Caledonian crow would use the stick tool is (0.8611, 0.9714). The confidence interval obtained using the binomial distribution function is (0.8727, 0.9714). Ironically, the interval derived from the exact binomial distribution has only an approximate 95-percent confidence level.

```
setwd("C:\\Users\\R")
df1 <-read.csv("binomial summary data.csv")
# binomial summary data.csv has three columns:
# Y = no. of successes out of N
# N = sample size
# alpha = confidence coefficient (confidence = 1 - alpha)
#
# ref:
# Clopper, C.J., Pearson, E.S., (1934) The use of confidence or
#                           fiducial limits illustrated in the case of the binomial
#                           Biometrika, 26, pp. 404-413
#

pL <- c()
pU <- c()
p <- c()
attach(df1)
alpha = 0.05
p <- Y / N
nsets-<nrow(df1)

for (i in 1:nsets){
 if (Y[i] <= 0) {
#                   note that the case Y[i] < 0 is an absurdity
  pL[i] <-0
  }
 else {
  pL[i] <-(2*Y[i])*qf(alpha/2,2*Y[i],2*(N[i] - Y[i] + 1))/(2*(N[i]-Y[i]+1) +
(2*Y[i])*qf(alpha/2,2*Y[i],2*(N[i]-Y[i] + 1)))
  }

 if (X[i] >= N[i]) {
#                   note that the case Y[i] > N[i] is an absurdity
  pU[i] <-1
  }
 else {
  pU[i] <-(2*(Y[i]+1))*qf(1-alpha/2,2*(Y[i]+1),2*(N[i] -Y[i]))/(2*(N[i]-Y[i]) +
(2*(Y[i]+1))*qf(1-alpha/2,2*(Y[i]+1),2*(N[i] - Y[i])))
  }
 }

df2 <-data.frame(Y,N,p,pL,pU)
write.csv(df2,"binomial confidence limits.csv")
```

FIGURE 2.1
Clopper–Pearson confidence intervals in R.

TABLE 2.2

Clopper–Pearson Confidence Intervals

Test	Y	N	p	pL	pU
1	17	20	0.85	0.621073	0.967929
2	85	100	0.85	0.764693	0.913546

2 × 2 Tables

The R-function that computes the p-values for Fisher's exact test is *fisher. test*(). Its argument is a table or matrix. As an example, consider a hypothetical experiment to test the effect of egg color on egg rejection behavior in northern mockingbirds (*Mimus polyglottos*). Northern mockingbird nests are sometimes parasitized by brown-headed cowbirds (*Molothrus ater*), and it has been suggested that egg color might play a role in determining the rate of parasitic egg rejection in northern mockingbirds (Quinn and Tolson, 2009). Suppose that two different artificial eggs were placed in each of 20 mockingbird nests: One mimetic egg painted to look like a mockingbird egg, and one nonmimetic egg that was pure white. Suppose that the researchers then record whether each egg is rejected by the host ($X = 1$) or not ($X = 0$). Suppose that 17 individuals rejected the nonmimetic egg, and 10 rejected the mimetic egg. Figure 2.2 shows some R code for computing the table. Relating these results to the generic 2 × 2 Table 2.1, N1 = N2 = 20, X1 = 17, and X2 = 10.

The odds ratio is a very useful concept when comparing probabilities. More will be said about odds and odds ratios in the chapter on logistic

```
#
# Xeq1 represents the condition that X = 1 (reaction)
# Xeq0 represents the condition that X = 0 (no reaction)
Xeq1 <- c(17,10) # Xeq1 shows the numbers of individuals where X = 1 for group #1 (stranger)
and # group 2 (familiar), respectively
Xeq0 <- c(3,10) # Xeq0 shows the numbers of individuals where X = 0 for group #1 and # group
2, respectively
cont.table <- cbind(Xeq1,Xeq0) # this is the 2 x 2 table
#
# this function computes Fisher's Exact Test for a 2x2 table
#
fisher.test(cont.table)

If you run this code, you will find the output to be:

Fisher's Exact Test for Count Data

data:   cont.table
p-value = 0.04074
alternative hypothesis: true odds ratio is not equal to 1
95 percent confidence interval:
 1.058679 37.982868
sample estimates:
odds ratio
 5.410903
```

FIGURE 2.2
R-code for Fisher's exact test.

FIGURE 2.3
New Caledonian crow (*Corvus moneduloides*).

regression, and the generalization of the 2 × 2 table to R × C tables (i.e., R rows and C columns).

Figure 2.3 is an illustration of a New Caledonian crow using a stick tool.

Theoretical Aspects

Single Proportion

Using the 0/1 valuation for the Bernoulli variables:

$$Y = \sum_{i=1}^{n} X_i$$

The probability of getting a specific value of Y, call it y, is the binomial mass function:

$$Pr\{Y = y\} = \binom{n}{y} p^y (1-p)^{n-y}$$

The cumulative distribution function for the binomial is

$$F_Y(y) = Pr\{Y \le y\} = \sum_{k=0}^{y} \binom{n}{k} p^k (1-p)^{n-k}$$

Although n is technically a parameter, it does not require estimation. However, p is generally unknown, so it must be estimated from the data. The maximum likelihood estimate of p, using a sample of Bernoulli values $x_1, x_2, \ldots, x_n$, is

$$\hat{p} = \frac{1}{n}\sum_{i=1}^{n} x_i$$

This is just the arithmetic average of the sample results.

Once data are obtained and an estimate of p is computed, we would like to get a confidence interval for the parameter. Suppose that y out of n observations had the characteristic of interest (i.e., y of the x_i values were equal to one). One way to get a confidence interval for p is to find two values, p_L and p_U, such that

$$Pr\left\{Y \le y \middle| p_L\right\} \approx 1 - \frac{\alpha}{2} \approx \sum_{k=0}^{y} \binom{n}{k} p_L^k (1-p_L)^{n-k}$$

and

$$Pr\left\{Y \le y \middle| p_U\right\} \approx \frac{\alpha}{2} \approx \sum_{k=0}^{y} \binom{n}{k} p_U^k (1-p_U)^{n-k}$$

Continuing with the first two examples presented, if $n = 20$ and $y = 17$, then

$$Pr\left\{Y \le y \middle| p_L = 0.683\right\} = \sum_{k=0}^{17} \binom{20}{k} p_L^k (1-p_L)^{20-k} \approx 0.975$$

and

$$Pr\left\{Y \le y \middle| p_U = 0.968\right\} = \sum_{k=0}^{17} \binom{20}{k} p_L^k (1-p_L)^{20-k} \approx 0.025$$

Thus the (approximately) 95-percent confidence interval for $p = Pr\{X = 1\}$ (the Bernoulli parameter) is (0.683, 0.968). In other words, if the probability of observing the characteristic of interest is about 0.683, then there is a fairly high chance that out of $n = 20$ individual observations, 17 or fewer such characteristics would be observed. Conversely, if the probability of the characteristic was as high as 0.968, then there is a fairly low chance that 17 or fewer such characteristics would be observed out of $n = 20$ individuals. All told, with this sample, we can only say we are nearly certain that the truth lies somewhere in the interval (0.683, 0.968).

Note that 17 out of 20 is exactly 85 percent. Now suppose we had observed $n = 100$ individuals, and 85 of them exhibited the characteristic of interest. Then

$$Pr\left\{Y \le y \middle| p_L = 0.777\right\} = \sum_{k=0}^{85} \binom{100}{k} p_L^k (1-p_L)^{100-k} \approx 0.974$$

and

$$Pr\left\{Y \le y \middle| p_U = 0.913\right\} = \sum_{k=0}^{85} \binom{100}{k} p_L^k (1-p_L)^{100-k} \approx 0.026$$

So, with $y = 85$ and $n = 100$, our approximate 95-percent confidence interval is now (0.777, 0.913). There are two important things to notice. First, since the observations are discrete (either the characteristic is there or not) the confidence level is not exactly 95 percent. Secondly, although in both cases, 85 percent of the sample was observed to have the characteristic, the confidence interval was wider with the smaller sample size. The R function *pbinom()* was used to compute the confidence intervals. Specifically,

pbinom(q=17,size=20,prob=0.683,lower.tail=TRUE) ≈ 0.975
pbinom(q=17,size=20,prob=0.968,lower.tail=TRUE) ≈ 0.025

The default value of the lower.tail parameter is TRUE. Unlike the case of continuous random variables, for discrete variables, $Pr\{Y = y\}$ is not necessarily equal to 0. Whereas when lower.tail = TRUE, the pbinom function returns $Pr\{Y \le y\}$, when lower.tail=FALSE, it returns $Pr\{Y > y\}$ (i.e., strictly greater than).

The reader can decide whether the Clopper–Pearson intervals are "close" to those obtained using the binomial distribution. The advantages of Clopper–Pearson are that the confidence level is "exactly" 95 percent ($100(1 - \alpha)\%$), and they are easier to compute (i.e., a simple, closed-form expression as opposed to derived via an iterative procedure).

2 × 2 Tables

Fisher's exact test is based on the condition that the row and column totals are fixed. If X1 and X2 are the observed numbers of individuals having the characteristic of interest ($X = 1$), then the probability of observing the table is given by

$$p_{table} = \frac{\binom{X1+X2}{X1}\binom{N1-X1+N2-X2}{N1-X1}}{\binom{N1+N2}{N1}}$$

Fisher's test computes all the probabilities for any table where the row totals are N1 and N2, and the column totals are X1 + X2 and N1 + N2 − (X1 + X2).

Key Points for Chapter 2

- Binary random variables are those that have only two possible values; they are connected to observations that can be described as either "event occurred" or "event did not occur," or "characteristic present" or "characteristic not present."
- The outcome of interest is generally coded as "1" and its negation is coded as "0."
- A Bernoulli random variable is a binary variable with a fixed probability of success, or in other words $Pr\{X = 1\} = p$.
- Binomial variables are sums of a set of n independent Bernoulli variables.
- Tests of hypotheses and confidence intervals for the parameter p involve the binomial distribution or approximations to the binomial.
- Comparing two groups using a binomial variable may be performed using Fisher's exact test.

Exercises and Questions

1. In a sample of $n = 30$ blue crayfish (*Procambarus alleni*), 12 of them attacked a minnow decoy. Construct a 95-percent confidence interval for the probability of attack.

2. In a sample of $n = 40$ female seed beetles (*Callosobruchus maculatus*), 22 were virgins and 18 were mated. Of the virgins, 15 chose to mate when presented with a male. Of the mated females, 11 chose to remate. Is there a significant difference between the virgins and mated females in the propensity to mate?

3

Continuous Variables

General Ideas

Continuous variables are those measurements or observations for which computing a mean or a standard deviation makes sense. These are variables that can be expressed with decimal numbers. Count data are technically not continuous. In some cases, we can treat count data as though they were continuous. Most of the discussion in this chapter will be about means of continuous variables, although some of it can be applied to other parameters, such as standard deviations. Tests will include one sample *t*-tests, two sample *t*-tests (with or without assumptions of equal variances), rank-based tests, and permutation tests. The *t*-test that does not rely on the assumption of equal variances is called Welch's test. The rank-based test we will discuss is called the Mann–Whitney–Wilcoxon test.

Rank-based tests and permutation tests require some explanation. In rank-based tests, the data are replaced with their respective ranks, where they are sorted from smallest to largest, and their numerical order is the "rank." The tests are then done on the ranks. This reduces the influence of values that depart greatly from the central tendency (i.e., the value with rank n is only one rank away from the next largest value in the sample), and the standard deviations of ranks tend to be less disparate. As mentioned earlier, the rank-based two sample test is called the Mann–Whitney–Wilcoxon test (Conover, 1980).

In permutation tests (Good, 1994), rather than relying on distributional assumptions, a sort of empirical null distribution for the test statistic is derived by randomly sampling the actual data and then randomly assigning the values to groups. Once a random assignment is performed, the test statistic is computed for the "permuted" sample, and the random sampling is done again. This process iterates for some "large" number of times, and the test statistic as computed with the "unpermuted" data is compared to this permutation distribution. If the test statistic falls in the lower or upper tails of the permutation distribution, then the null hypothesis is rejected.

Another technique related to permutation tests is bootstrapping (Efron, 1982). Bootstrapping will be discussed in the context of computing confidence intervals.

Examples with R Code

The R function *t.test*() is used to perform tests of hypotheses concerning means for either one or two samples. For a one sample test, the data are in the form of a column vector, with values of the response variable forming the column. If the data are stored in a column vector:

response $< -c$(30.0, 31.0, 29.9, 28.8, 32.3, 33.6, 29.7, 30.1, 30.5, 32.0)

then to test the hypotheses

$$H_0: \mu = 30$$
$$H_1: \mu \neq 30$$

use *t.test*(response,mu=30). Figure 3.1 shows the R commands and associated output.

The degrees of freedom (df) is equal to 9, or $n - 1$. Since the *p*-value is 0.118, the null is not rejected at the $100\alpha\% = 5\%$ level. Also, the null value of μ (30) is contained within the 95-percent confidence interval. Thus the null is not rejected.

A confidence interval for standard deviation, σ, can be computed using the Chi-squared distribution. The R function *qchisq*() is used. Figure 3.2 shows how the confidence interval for the standard deviation, using the data above, is computed.

So, if we wanted to test the hypothesis that the standard deviation, σ, was equal to 1.0, we would see if 1.0 fell in the interval (s.low, s.upp) $\approx$ (1.048, 2.782), which it does not. So, the hypothesis that $\sigma = 1.0$ is rejected.

An alternative to using the Student's *t* or chi-squared distributions for hypothesis tests or computing confidence intervals is called *bootstrap sampling* (Efron, 1982). To perform bootstrap sampling, the data are randomly

```
> response <-c(30.0, 31.0, 29.9, 28.8, 32.3, 33.6, 29.7, 30.1,30.5, 32.0)
> t.test(response,mu=30) #the null hypothesis is that the population mean = 30

OUTPUT:
     One Sample t-test
data:  response
t = 1.7281, df = 9, p-value = 0.118
alternative hypothesis: true mean is not equal to 30
95 percent confidence interval:
 29.75585 31.82415
sample estimates:
 mean of x
30.79
```

FIGURE 3.1
A one-sample *t* test.

```
> response <-c(30.0, 31.0, 29.9, 28.8, 32.3, 33.6, 29.7, 30.1,30.5, 32.0)
> sigest <-sd(response)
> n <-length(response)
> s.low <-sigest*sqrt(n/qchisq(p=0.975,df=n-1))
> s.upp <-sigest*sqrt(n/qchisq(p=0.025,df=n-1))
> sigest
[1] 1.445645
> s.low
[1] 1.048153
> s.upp
[1] 2.781944
```

FIGURE 3.2
Confidence interval for σ.

sampled, with replacement, and for each random resampling, the statistic of interest is computed. After a "large" number of such resampled statistics are compiled, the $100(\alpha/2)$th and $100(1 - \alpha/2)$th percentiles of the resampled statistics' distribution are used to represent the $100(1 - \alpha)$% confidence interval.

Figure 3.3 shows R code for computing bootstrap confidence intervals for μ and σ.

The 95-percent confidence interval for the mean was (29.96, 31.66). Compare this to the conventional Student's t based interval of (29.76, 31.82).

Figure 3.4 and Figure 3.5 show histograms of the bootstraps distributions for mean and ds, respectively.

So far the discussion has been about single samples. Often two samples, observed or measured under two "conditions" or perhaps two distinct groups of individuals, are compared. To perform a two-sample test, there need to be two variables in the dataset. One is the response variable (response), and the other is a variable that indicates the group to which an observation belongs (group). For a two-sample test, the specification for the R function would be

$$t.test(\text{response} \sim \text{group})$$

The null distribution of the test statistic is Student's t with $n_1 + n_2 - 2$ degrees of freedom, provided that the populations or data-generating processes from which the two samples come have identical standard deviations, σ. Even if the means are in fact identical, the violation of equal standard deviations (or variances) can alter the p-value outcome. There are several possible remedies. The one employed by R is called Welch's test, which uses a null distribution having degrees of freedom that are weighted by the sample standard deviations from the two samples. The $t.test()$ function actually defaults to Welch's test, unless the input parameter var.equal is set to TRUE. Table 3.1 shows some data formatted for the two sample $t.test$ function. Figure 3.6 shows some R code that executes the function using both var.equal=TRUE and var.equal=FALSE, together with the output shown in Figure 3.7. In particular, notice that the degrees of freedom for var.equal=TRUE is in fact $n_1 + n_2 - 2 = 20 + 20 - 2 = 38$, whereas for var.equal=FALSE, the degrees of freedom are 31.941. In this particular case, the difference in p-values was very small, and both tests were significant at the $\alpha = 0.05$ level.

```
setwd("C:\\Users\\Statistical Methods for Animal Behavior Field Studies\\Statistical Data &
Programs")
#
#
response <-c(30.0, 31.0, 29.9, 28.8, 32.3, 33.6, 29.7, 30.1,30.5, 32.0)
boot.sample <-c()
boot.mean <-c()
boot.sd <-c()

alpha <-0.05

mean.estimate <-mean(response)
sd.estimate <-sd(response)


nreps <-2000
nsize <-length(response)

#
#  The Boostrap Loop:
#
for (i in 1:nreps) {
 boot.sample <-sample(response,size=nsize,replace=TRUE)
 boot.mean[i] <-mean(boot.sample)
 boot.sd[i] <-sd(boot.sample)
 }
lcl.mean <-quantile(boot.mean,probs=c(alpha/2))
ucl.mean <-quantile(boot.mean,probs=c(1-alpha/2))
lcl.sd <-quantile(boot.sd,probs=c(alpha/2))
ucl.sd <-quantile(boot.sd,probs=c(1-alpha/2))

upper.mean <-max(boot.mean)
lower.mean <-min(boot.mean)
hist(boot.mean,xlim=c(lower.mean-1.5,upper.mean+1.5),freq=F)#this plots a histogram of the
bootstrap mean distribution

dev.new() # this allows for a second plot to be created without overwriting the first
#
#
#
upper.sd <-max(boot.sd)
lower.sd <-min(boot.sd)
hist(boot.sd,xlim=c(lower.sd-1.5,upper.sd+1.5),freq=F)#this plots a histogram of the bootstrap sd
distribution

lcl.mean
ucl.mean
lcl.sd
ucl.sd

OUTPUT:
> lcl.mean
 2.5%
29.96
>ucl.mean
97.5%
31.66
> lcl.sd
  2.5%
0.6745945
> ucl.sd
 97.5%
1.845457
```

FIGURE 3.3
Bootstrap sampling R code with output.

An R function that computes the rank-based test is called *wilcox.test*(), which is in the R package MASS. The code in Figure 3.6 can be easily modified by adding:

$$wilcox.test(\text{response} \sim \text{group})$$

The output from this function is given in Figure 3.8.

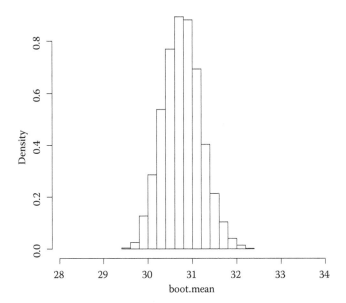

FIGURE 3.4
Histogram of bootstrap mean distribution.

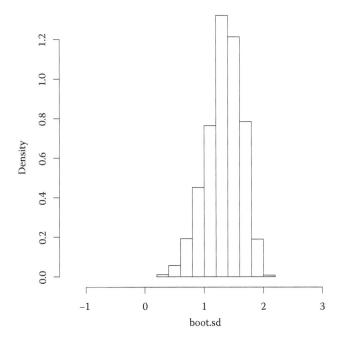

FIGURE 3.5
Histogram of bootstrap SD distribution.

TABLE 3.1

Some Two-Sample Data

Response	Group
27.8	1
30.3	1
34.6	1
26.4	1
30.6	1
29.6	1
28.8	1
28.7	1
29	1
29.4	1
29.7	1
31.6	1
31.1	1
30.3	1
27.2	1
28.5	1
32	1
25.9	1
30.3	1
29.8	1
27.9	2
34.4	2
37.8	2
27.5	2
31.1	2
31.8	2
29.6	2
29.3	2
33.5	2
32.5	2
33.5	2
38.5	2
27.8	2
36.2	2
30.8	2
30.1	2
34.8	2
34.8	2
31.7	2
32.8	2

```
setwd("C:\\Users\\Statistical Methods for Animal Behavior Field Studies\\Statistical Data
& Programs")
df1 <-read.csv("20160916 Example 3.2 Two Sample Test.csv")
#
#
# VARIABLES:
#  response
#  group
#
attach(df1)
t.test(response ~ group,var.equal=TRUE)
t.test(response ~ group,var.equal=FALSE)
----------------------------------------------------------------------------
```

FIGURE 3.6
R-code for two-sample t.test.

```
Two Sample t-test
data:  response by group
t = -3.2653, df = 38, p-value = 0.002319
alternative hypothesis: true difference in means is not equal to 0
95 percent confidence interval:
 -4.438702 -1.041298
sample estimates:
mean in group 1 mean in group 2
     29.58          32.32

> t.test(response ~ group,var.equal=FALSE)

    Welch Two Sample t-test

data:  response by group
t = -3.2653, df = 31.941, p-value = 0.002611
alternative hypothesis: true difference in means is not equal to 0
95 percent confidence interval:
 -4.449348 -1.030652
sample estimates:
 mean in group 1 mean in group 2
29.58          32.32
```

FIGURE 3.7
Output for two-sample t.test.

```
> wilcox.test(response ~ group)


    Wilcoxon rank sum test with continuity correction


data:  response by group

W = 96.5, p-value = 0.005314

alternative hypothesis: true location shift is not equal to 0
```

FIGURE 3.8
Output for two-sample Mann–Whitney–Wilcoxon test (wilcox.test).

```
setwd("C:\\Users\\Statistical Methods for Animal Behavior Field Studies\\Statistical Data &

Programs")

df1 <-read.csv("20160916 Example 3.2 Two Sample Test.csv")
#
#
# VARIABLES:
#  response
#  group
#
tnull <-c()
attach(df1)
alpha <-0.05

mean.group1 <-mean(response[group==1])
mean.group2 <-mean(response[group==2])
sd.group1 <-sd(response[group==1])
sd.group2 <-sd(response[group==2])

toriginal <-t.test(response ~ group,var.equal=FALSE)
t.star <-toriginal$statistic

nreps <-1000
nsize <-nrow(df1)

#
#  The permutation Loop:
#
for (i in 1:nreps) {

 rand <-runif(nsize,0,1) #obtains random numbers between 0 and 1
 permute.data <-response[order(rand)] # this permutes the response data
 df.perm <-data.frame(group,permute.data) #this creates a permuted dataframe
 t.perm <-t.test(df.perm$permute.data ~ df.perm$group)
 tnull[i] <-t.perm$statistic
 }

t.low <-quantile(tnull,probs=c(alpha/2))
t.high <-quantile(tnull,probs=c(1-(alpha/2)))

pos.star <-abs(t.star)
neg.star <--1*t.star
```

FIGURE 3.9

R Code for permutation test to compare means of two groups.   (*Continued*)

Figure 3.9 shows R code for computing a permutation "*t* test." Figure 3.10 shows the output. The number of iterations, nreps, was set to 1000; this was a somewhat arbitrary value.

The variable p.comp is the *p*-value based on the permutation null distribution.

Figures 3.11 and 3.12 show, respectively, the histogram and cumulative distribution function (CDF) for the null distribution of the t statistic (tnull). Notice that the original unpermuted test statistic (t.star) is in the far left tail of the null distribution.

In all cases for this particular example, the null hypothesis of equality of means was rejected. Of course, these methods may not always agree. One of the primary advantages of permutation tests and bootstrap sampling is that no assumptions need to be made about the distributional characteristics of the data in order to be able to make inferences. While rank-based methods reduce the reliance on assumptions, they often require "large sample" approximations to provide inferences.

```
p.low <-sum(tnull <= neg.star)
p.med <-sum(tnull <= t.star) #no. permuted  results <= t.star
p.high <-sum(tnull >= pos.star) #no. permuted results >= t.star

if (t.star <= 0.0)  p.comp <-(p.med + p.high) else p.comp <-(p.low + p.high)

p.comp <-p.comp / nreps #permutation test p-value

pos.star
neg.star
mean.group1
sd.group1
mean.group2
sd.group2

t.star
t.low
t.high
p.low
p.med
p.high
p.comp

upper <-max(tnull)
lower <-min(tnull)
hist(tnull,xlim=c(lower-1.5,upper+1.5),freq=F)#this plots a histogram of the permutation null

sorted.tnull <-sort(tnull)
dev.new()# this allows for a second plot to be created without overwriting the first
#
# this plots the cumulative distribution function for tnull
# extended to go beyond the range of the values in tnull
#

plot(sorted.tnull, main="CDF for tnull",xlab="t
values",ylab="Pr{tnull<=t}",(1:nreps)/nreps,type="s",xlim=c(lower-1.5,upper+1.5),ylim=c(0,1))
#the y-axis is (1:nreps)/nreps
```

FIGURE 3.9 (CONTINUED)
R Code for permutation test to compare means of two groups.

Theoretical Aspects

We begin with a single sample of values from a normally distributed data-generating process. If the observed values of random variable X are $x_1, x_2, ..., x_n$, then we know the maximum likelihood estimators for the parameters μ and σ^2 are

$$\hat{\mu} = \frac{1}{n}\sum_{i=1}^{n} x_i$$

and

$$\hat{\sigma}^2 = \frac{1}{n}\sum_{i=1}^{n} (x_i - \hat{\mu})^2$$

We have seen how to create a $100(1 - \alpha)\%$ confidence interval for the parameter μ:

```
> pos.star
     t
3.265341
> neg.star
     t
3.265341
> mean.group1
[1] 29.58
> sd.group1
[1] 1.993569
> mean.group2
[1] 32.32
> sd.group2
[1] 3.179308
>
> t.star
     t
-3.265341
> t.low
   2.5%
-2.217032
> t.high
  97.5%
2.017223
> p.low
[1] 995
> p.med
[1] 3
> p.high
[1] 5
> p.comp
[1] 0.008
```

FIGURE 3.10
Output of permutation "*t* test."

$$(\mu_{low}, \mu_{high}) = \left(\hat{\mu} - t_n \frac{\hat{\sigma}}{\sqrt{n}}, \quad \hat{\mu} + t_n \frac{\hat{\sigma}}{\sqrt{n}} \right)$$

Where t_n is the $100(1 - \alpha/2)^{th}$ percentile of a Student's t distribution with n degrees of freedom. Keep in mind that we are using the maximum likelihood estimator for σ, and not the unbiased estimator, which has $n - 1$ in the denominator instead of n. The R function sd() computes the unbiased estimator, so use $n - 1$ degrees of freedom when sd() is used.

If one were performing a hypothesis test for the mean, μ, such as

$$H_0: \mu = \mu_0$$
$$H_1: \mu \neq \mu_0$$

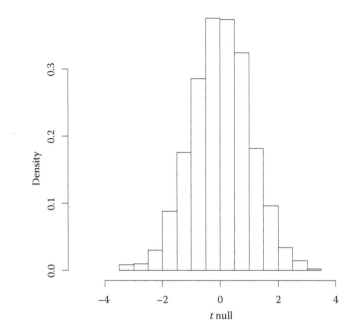

FIGURE 3.11
Histogram of permutation "null" distribution.

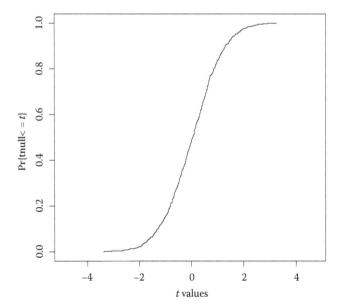

FIGURE 3.12
Cumulative distribution function of permutation "null" distribution.

Then compute the test statistic:

$$t = \frac{\hat{\mu} - \mu_0}{\hat{\sigma} \big/ \sqrt{n}}$$

And compare it to $\pm t_n$, the $\pm 100(1 - \alpha/2)^{\text{th}}$ percentile of a Student's t distribution with n degrees of freedom. If $t < -t_n$ or $t > +t_n$, then reject the null hypothesis.

A $100(1 - \alpha)$% confidence interval for σ can also easily be constructed, with the formula:

$$(\sigma_{low}, \sigma_{high}) = \left(\hat{\sigma} \sqrt{\frac{n}{\chi^2_{1-\frac{\alpha}{2}}(n)}}, \ \hat{\sigma} \sqrt{\frac{n}{\chi^2_{\frac{\alpha}{2}}(n)}} \right)$$

$\chi^2_p(n)$ is the $100p^{\text{th}}$ percentile of a chi-squared distribution having n degrees of freedom.

If the estimate of σ^2 is the unbiased estimator:

$$\hat{\sigma}^2 = \frac{1}{n-1} \sum_{i=1}^{n} (x_i - \hat{\mu})^2$$

Then use $\chi^2_p(n-1)$, the $100p^{\text{th}}$ percentile of a chi-squared distribution having $n - 1$ degrees of freedom in the formula.

One-sample tests are fairly simple. In two-sample tests, the objective is to decide whether the parameter of interest for the populations or data-generating processes from which the two samples came is actually the same or not. The ideas for inference and confidence intervals are very similar to those for single samples. Suppose there are two "groups" of individuals, group 1 and group 2. The grouping can be based on a natural phenomenon, like geographic location, or on an experimental phenomenon, such as exposure to two different stimuli. The first hypothesis we will entertain concerns the means of some continuously-valued response variable. Namely, consider:

$$H_0: \mu_1 = \mu_2$$
$$H_1: \mu_1 \neq \mu_2$$

Here we are not attempting to show that the average response of one particular group (say group 1) is *larger* than the average response for the other group. Rather, we just want to know if the means differ. Assume that the sample sizes for the two groups are n_1 and n_2 , respectively. As in the case of the single sample, we will form a test statistic:

$$t = \frac{\hat{\mu}_1 - \hat{\mu}_2 - 0}{\sqrt{\dfrac{\hat{\sigma}_1}{n_1} + \dfrac{\hat{\sigma}_2}{n_2}}}$$

This statistic, given that the null hypothesis is true, would have a Student's t distribution with $n_1 + n_2$ degrees of freedom (or $n_1 + n_2 - 2$ degrees of freedom if the unbiased estimates of σ are used). Subtracting 0 in the numerator is just to emphasize the fact that the null hypothesis is that the difference in the two means is 0. One of the assumptions that makes the test statistic have its Student's t distribution is that the variances for the two groups are actually equal. Of course, the sample estimates may differ, but the assumption is about the population, or data-generating process.

The test statistic for the Mann–Whitney–Wilcoxon test is the normalized sum of ranks for either one of the two groups. That is, if $r_{ij} = rank(x_{ij})$ is the rank of the observed or measured value xi in group j, and

$$S_k = \sum_{i=1}^{n_k} r_{ik}$$

(for $k = 1$ or 2)

and

$$N = n_1 + n_2$$

then the test statistic is

$$W = \frac{S_k - n_k \dfrac{N+1}{2}}{\sqrt{\dfrac{n_1 n_2}{N(N-1)} \displaystyle\sum_{j=1}^{2} \sum_{i=1}^{n_j} r_{ij}^2 - \dfrac{n_1 n_2 (N+1)^2}{4(N-1)}}}$$

If the null hypothesis is true, then W has approximately a normal distribution with mean 0 and standard deviation 1 (called a standard normal). The critical values are the $100(\alpha/2)^{th}$ and $100(1 - \alpha/2)^{th}$ percentiles of the standard normal. Conversely, if $\Phi(z)$ is the cumulative distribution function for the standard normal, and $\Phi^{-1}(p)$ is the 100th percentile of the standard normal, then the critical values would be

$$z_{\frac{\alpha}{2}} = \Phi^{-1}\left(\frac{\alpha}{2}\right)$$

and

$$z_{1-\frac{\alpha}{2}} = \Phi^{-1}\left(1-\frac{\alpha}{2}\right)$$

The p-value for W would be

$$p = 2\left(1 - \Phi(|W|)\right)$$

In the case where there are no "ties" in the data (i.e., no two values in the entire dataset are the same to whatever numerical precision the measurements are expressed), then the function *wilcox.test* will compute an "exact" p-value based only on the sum of the ranks, S_k.

One of the advantages of rank-based methods is that they do not generally rely on the need for data to come from normally distributed populations or data-generating processes. The strength and the weakness of rank-based methods is that disparity between the extreme ranks (lowest and highest, for example) and the center rank is not affected by the magnitude of the extreme values in natural units. Thus, the effects of extreme values on the test statistic are reduced. In fact, when the data come from a highly asymmetrical distribution, and the difference between mean and median is greater, then rank-based methods can in a certain sense fail to detect meaningful differences in average values.

Permutation tests are the collective name for another set of methodologies that also dispense with the necessity of distributional assumptions in order to derive "null distributions" for test statistics. The permutation test methodology is based on random reassignment of data to groups. First, the test statistic is computed in the same way as in the cases of parametric tests. However, in order to determine a critical region, the data are iteratively resampled, randomly assigned to the groups, and the test statistic is recomputed. The notion is that if the groups really have no effect on the response, then assigning the results to any group would not affect the difference in, say the mean value, of the groups. After random sampling and assignment many times, a "null distribution" for the test statistic is formed, and the original test statistic is compared to it. If the original statistic falls in, say, the $100(\alpha/2)^{th}$ percentile or the $100(1 - \alpha/2)^{th}$ percentile, then the null is rejected. Figure 3.13 shows the flow of a permutation test process.

```
Compute test statistic, t*

For i = 1 to N:

 Randomly sample the data, assigning values to groups at random

 (but preserving sample size for each group)

 Compute a permuted test statistic, tp

 Store the permuted test statistic

 Loop

Compare original test statistic, t*, to permutation distribution

If t* falls in the 100(α/2)% or 100(1-α/2)% tails of the permutation distribution, then

 Reject the null hypothesis

Else

 Do not reject the null hypothesis
```

FIGURE 3.13
Permutation test process flow.

Compute test statistic, t*
For i = 1 to N:

Randomly sample the data, assigning values to groups at random (but preserving sample size for each group)

Compute a permuted test statistic, t_p

Store the permuted test statistic

Loop

Compare original test statistic, t*, to permutation distribution

If t* falls in the $100(\alpha/2)$% or $100(1 - \alpha/2)$% tails of the permutation distribution, then

Reject the null hypothesis

Else

Do not reject the null hypothesis

Key Points for Chapter 3

- Confidence intervals provide a "feasible range" of values for a parameter.
- Parametric methods usually require making assumptions about the distributional characteristics of populations or data-generating processes; these assumptions are often unverifiable.

- Rank-based methods aid in reducing the reliance on assumptions; often, however, approximations are made to compute p-values.
- Permutation tests take data and randomly reassign them to groups or treatments in order to create a "null" distribution for a statistic without the need to make restricting assumptions.
- Bootstrapping resamples the data withpout any reassignment, in order to obtain a better estimate of some parameter, and compute an empirical confidence interval.
- Permutation tests and bootstrapping require the fewest number of assumptions for inference; they may not be convenient for computing purposes.

Exercises and Questions

1. The usual test statistic for comparing two standard deviations is the ratio of the two sample variances:

$$F = \frac{S_1^2}{S_2^2}$$

Under the assumption that the data for each group come from a normally distributed population, under the null hypothesis of equal standard deviations for the two groups this ratio has a F distribution with $n_1 - 1$ and $n_2 - 1$ degrees of freedom (or n_1 and n_2 degrees of freedom, if the maximum likelihood estimator for variance is used instead of the unbiased estimator).

Construct a permutation test for the Example 3.2 data to compare the standard deviations of the two groups; compare the conclusions drawn using the "conventional" F test; use $\alpha = 0.05$.

2. Construct two-sided 95-percent confidence intervals for each group's (Example 3.2 data) standard deviation; use a bootstrap approach. Recall that the conventional two-sided confidence interval for the standard deviation is

$$\left(\frac{\sqrt{n-1}S}{\sqrt{\chi_{1-\frac{\alpha}{2}}^2(n-1)}}, \frac{\sqrt{n-1}S}{\sqrt{\chi_{\frac{\alpha}{2}}^2(n-1)}} \right)$$

and $\chi_p^2(n-1)$ is the 100th percentile of a chi-squared distribution with $n-1$ degrees of freedom.

4

The Linear Model: Continuous Variables

General Ideas

A multiple linear regression model can be expressed in the following fashion:

$$Y = \beta_0 + \beta_1 Z_1 + \beta_2 Z_2 + \ldots + \beta_k Z_k + \varepsilon$$

Y is called the response variable, with a random component ε. We would like to be able to predict the value of Y, at least on the average, given values for $Z_1, Z_2, \ldots, Z_k$. The values of the parameters β_j, $j = 0, k$, are in general unknown, and must be estimated using data.

The variables Z_j, $j = 1, k$ are called regressors or predictors. They can in turn be functions of other variables, or nonlinear functions of the other Z variables. What they cannot be are linear functions of the other Z variables. So, for example, suppose that X represents a continuous variable that is directly measured or observed on an individual. Then it is permissible for

$$Z = X$$
$$Z = \sin(2\pi X)$$
$$Z = X^7$$

What is *not* permissible is

$$Z_1 = aZ_2 + bZ_3$$

When the regressors are linear functions of each other, they are said to be collinear. Collinearity makes it impossible to estimate the coefficients, β_j, when a linear model is employed. Initially, we will treat the Z_j as "fixed effects" regressors, meaning that their values, while varying between individuals, would not vary in any substantial way within an individual, at least within the conditions in which the individuals were observed.

The method of least squares is most commonly used to estimate the coefficients in a multiple linear regression.

TABLE 4.1

Form of a Multiple Linear Regression Dataset

Y	Z1	Z2	...	Zk
y1	z11	z12	...	z1k
y2	z21	z22	...	z2k
.	.	.	.	.
.	.	.	.	.
.	.	.	.	.
yn	zn1	zn2	...	znk

If n individuals are observed, and on each individual the values of Y and the Zj are measured or observed, the dataset could be represented as shown in Table 4.1.

The assumptions used for fitting and inference are that the noise, ε, has mean 0 and a constant standard deviation, that is, the standard deviation of noise does not vary with different values of the regressors. Of course, it is assumed that the regressors are not collinear.

Examples with R Code

Several years ago, we investigated whether Eastern gray squirrels (*Sciurus carolinensis*) use their tails to signal aggressive intent (Pardo et al., 2014). The following example with simulated data is based on a modified version of our study. We observed dyadic (pairwise) interactions between a dominant squirrel and a subordinate squirrel. Every time the dominant squirrel was aggressive toward the subordinate squirrel, we measured its degree of aggressiveness, and also recorded two variables describing the position of each squirrel's tail: the angle at which the squirrels bent their tails ("Curve"), and the point along the tail at which the bend occurred ("Bent"). For the purposes of this example, suppose that we observed a sample of $n = 35$ interactions. Further suppose that aggression was measured as the distance that the dominant squirrel traveled toward the subordinate squirrel during the encounter (continuous variable), and that "Curve" and "Bent" were both measured as continuous variables with values ranging between zero and one. We then fit a linear model to the data. Figure 4.1 shows the R code for fitting the model, together with plots of the response variable (Aggression) plotted against each of the regressors and the residuals plotted against the predicted (fitted) values of the response.

Figure 4.2 shows the output in the R console window.

The column labeled "estimates" contains the least squares estimates of the model parameters. The "Std. Error" column has the standard errors. The "t value" column has the t statistics used to test the hypotheses that the parameters are

```
setwd("C:\\Users\\Statistical Methods for Animal Behavior Field Studies\\Statistical Data &
Programs")
df1 <-read.csv("20161005 Example 4.1 Squirrel Aggression Dominant.csv")
#
#
# VARIABLES:
# Aggression
# D.Curve
# D.Bent
# S.Curve
# S.Bent
#
attach(df1)

linreg <-lm(Aggression ~ D.Curve + D.Bent + S.Curve + S.Bent)
#
# Make some plots:
#
# dev.new() allows more plots to be made without overwriting previous plots
#
plot(D.Curve,Aggression)
dev.new()
plot(D.Bent,Aggression)
dev.new()
plot(S.Curve,Aggression)
dev.new()
plot(S.Bent,Aggression)
dev.new()
plot(linreg$fitted.values,linreg$residuals)
abline(h=0.0)
summary(linreg)
```

FIGURE 4.1
R code for multiple linear regression.

```
Call:
lm(formula = Aggression ~ D.Curve + D.Bent + S.Curve + S.Bent)

Residuals:
    Min       1Q   Median       3Q      Max
-0.26539 -0.06665  0.01112  0.06166  0.17506

Coefficients:
              Estimate  Std. Error  t value  Pr(>|t|)
(Intercept)  -0.05205     0.06396   -0.814   0.422252
D.Curve       2.26550     0.08086   28.018    < 2e-16 ***
D.Bent        0.58607     0.11738    4.993   2.38e-05 ***
S.Curve       0.04362     0.07567    0.576   0.568599
S.Bent        0.52418     0.13083    4.006   0.000375 ***
---
Signif. codes:  0 '***' 0.001 '**' 0.01 '*' 0.05 '.' 0.1 ' ' 1

Residual standard error: 0.09709 on 30 degrees of freedom
Multiple R-squared:  0.9752,     Adjusted R-squared:  0.9719
F-statistic: 294.7 on 4 and 30 DF,  p-value: < 2.2e-16
```

FIGURE 4.2
Multiple regression output.

actually 0, and the Pr(>|t|) column has the corresponding *p*-values. The multiple R-squared is a measure of linearity between the response and all the regressors in a sort of aggregate sense. This value is always between zero and one; zero meaning no linear relationship at all, and one being a perfect linear relationship.

The residual standard error is actually the estimate of σ, the standard deviation of errors or noise.

Consider the plot of the residuals versus the predicted, or fitted, values, in Figure 4.3.

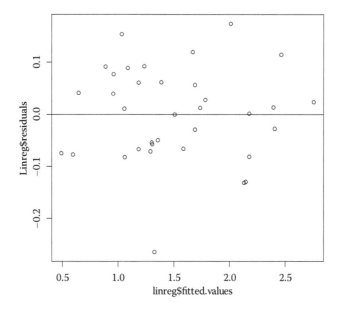

FIGURE 4.3
Squirrel tail position residuals versus fitted aggression values.

The hope is that if the assumptions necessary to allow for testing hypotheses about the model parameters, and of course for computing associated confidence intervals, are valid, that the residuals would be fairly symmetrically distributed around 0, regardless of the fitted value. This is a very informal check. However, with experience, intuition about the symmetry and model fit is often correct.

Now consider the output (Figure 4.4) of the R code of Figure 4.1, where the residuals have a variance that is proportional to the values of the regressors.

```
Call:
lm(formula = Aggression ~ D.Curve + D.Bent + S.Curve + S.Bent)

Residuals:
    Min      1Q   Median       3Q      Max
-0.86570 -0.32585 -0.05742  0.29933  1.02898

Coefficients:
             Estimate   Std. Error   t value    Pr(>|t|)
(Intercept)  0.411458    0.310059     1.327     0.194511
D.Curve      1.613163    0.391944     4.116     0.000278 ***
D.Bent       0.003337    0.568986     0.006     0.995359
S.Curve      0.265392    0.366822     0.723     0.474984
S.Bent      -0.069848    0.634193    -0.110     0.913034
---
Signif. codes:  0 '***' 0.001 '**' 0.01 '*' 0.05 '.'0.1 ' ' 1

Residual standard error: 0.4706 on 30 degrees of freedom
Multiple R-squared:  0.4076,     Adjusted R-squared:  0.3286
F-statistic:  5.16 on 4 and 30 DF,  p-value: 0.002764
```

FIGURE 4.4
Linear model output—proportional variance of residuals.

That is, the standard deviation (or variance) of ε is a constant multiplied by the value of some linear combination of regressors:

$$V[\varepsilon] = c(a_1 Z_1 + a_2 Z_2 + \ldots + a_k Z_k)$$

This is a fairly common and simple phenomenon that violates the constant standard deviation assumption.

Unlike the previous output shown in Figure 4.2, the only regressor that had a p-value less than 0.05 was D.Curve, even though the only thing that changed was the variance of the residuals (noise). The R^2 dropped from 0.9752 down to 0.4706. Figure 4.5 shows the plot of the residuals against the fitted values. In this figure, it appears that the dispersion of the residuals grows as the fitted values grow.

Figure 4.6 shows a plot of the original response, Aggression, plotted against the regressor D.Curve. Compare this to the "adulterated" response with the proportional variance, plotted against D.Curve in Figure 4.7.

It appears that the plot in Figure 4.6 has less dispersion than in Figure 4.7, although it is not too difficult to imagine a straight-line relationship between Aggression and D.Curve, even in Figure 4.7.

Now consider the plot of the original Aggression data against D.Bent in Figure 4.8, and the proportional variance version of the data plotted against the same regressor in Figure 4.9.

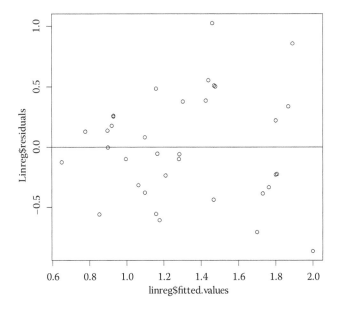

FIGURE 4.5
Squirrel tail position residuals versus fitted—proportional variance.

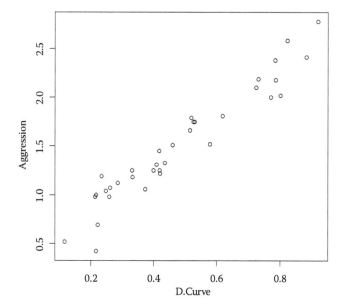

FIGURE 4.6
Aggression plotted against D.Curve—constant variance.

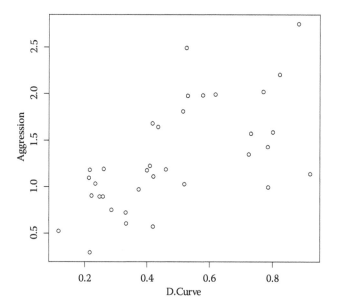

FIGURE 4.7
Aggression plotted against D.Curve—proportional variance.

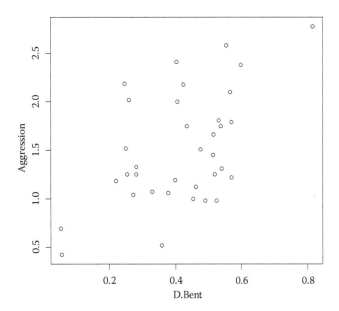

FIGURE 4.8
Aggression plotted against D.Bent—constant variance.

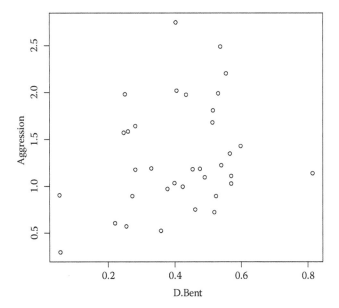

FIGURE 4.9
Aggression plotted against D.Bent—proportional variance.

It appears that in Figure 4.8 only a few observations may drive the significance of the p-value for the coefficient. Nevertheless, it does appear that some linear relationship exists. In Figure 4.9, the linearity is not as apparent, but perhaps not completely obliterated. Nevertheless, with the constant variance data, the p-value was approximately 0.0000285, and with proportional variance, the p-value was approximately 0.995.

The point of these illustrations is that assumptions of constant variance and normality are critical in the interpretation of p-values and measures such as R^2.

There are remedies for nonconstant variance of "errors." The response variable can be transformed and then the regression performed. One type of transformation is referred to as *Box–Cox*, named after its inventors (Box and Cox, 1964). If Y is the response, then the transformed response is

$$Y_T = \frac{Y^\lambda - 1}{\lambda}$$

where λ is a strategically chosen real number, other than 0. The transformation:

$$Y_T = \ln(Y)$$

is used when $\lambda \to 0$.

The regression is then performed on the transformed response. Predicted values can be "back-transformed" by inverting the transformation equation:

$$\hat{Y} = e^{\ln\left(\lambda \hat{Y}_T + 1\right)/\lambda}$$

Since there are random noise "errors" in the original response, as well as potential lack of fit, the inverted predictions will not match the observed "raw" response values, or predicted values using the untransformed response values perfectly.

The R function *boxcox()* in package MASS allows the analyst to find the "best" value of λ, by maximizing a likelihood function. In the squirrel example, the code to add is

```
boxtrans<-boxcox(object = linreg,plotit = TRUE)
```

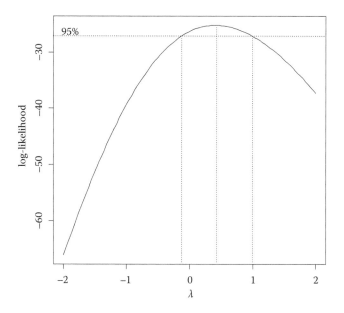

FIGURE 4.10
Box–Cox plot—squirrel data with proportional variance.

The object boxtrans contains potential values of λ and their corresponding (log) likelihood functions. The values of λ are stored in the attribute
boxtrans$x, and the log likelihood values in boxtrans$y. Choose the value of
λ = boxtrans$x that corresponds to the maximum value of boxtrans$y. The
plotit=TRUE parameter plots the log likelihood values against corresponding values of λ. Figure 4.10 shows the plot for the squirrel data with proportional variance.

Figure 4.11 shows the code using the proportional variance data, together
with *summary()* output for the transformed response. Figure 4.12 shows the
inverse-transformed predicted values plotted against the original untransformed predicted values.

One method for finding the position of the optimal value in boxtrans$x is
to use the *which.is.max()* function in package nnet.

A special kind of regressor variable is a cross-product of two or more
regressors. For example, if regressors Z_1 and Z_2 are included in a model,
then the term $Z_3 = Z_1^* Z_2$ can also be included. Since Z_3 is not a linear
function of the other regressors, including it does not induce collinearity in the model. Sometimes inclusion of cross-product terms, also called

```
setwd("C:\\Users\\SMFSBE\\Statistical Data & Programs")
df1<-read.csv("20161010 Example 4.2 Squirrel Aggression Dominant Proportional SD.csv")
#library(MASS) # package MASS has the boxcox() function
#library(nnet) # package nnet has the which.is.max() function#
#
# VARIABLES:
# Aggression
# D.Curve
# D.Bent
# S.Curve
# S.Bent
#
attach(df1)

linreg <-lm(Aggression ~ D.Curve + D.Bent + S.Curve + S.Bent)
#
# Make some plots:
#
# dev.new() allows more plots to be made without overwriting previous plots
#
plot(D.Curve,Aggression)
dev.new()
plot(D.Bent,Aggression)
dev.new()
plot(S.Curve,Aggression)
dev.new()
plot(S.Bent,Aggression)
dev.new()
plot(linreg$fitted.values,linreg$residuals)
abline(h=0.0)

summary(linreg)
dev.new()

boxtrans <-boxcox(object=linreg,plotit=TRUE)
lamdaopt <-boxtrans$x[which.is.max(boxtrans$y)] #this finds the optimal lamda

Aggtrans <-(Aggression**lamdaopt-1)/lamdaopt
lintrans <-lm(Aggtrans ~ D.Curve + D.Bent +
S.Curve + S.Bent)
summary(lintrans)
dev.new()
plot(lintrans$fitted.values,lintrans$residuals)
abline(h=0.0)
#

#
# The following code "inverts" the transformation for predicted values
#

invtrans <-exp(log(lamdaopt*lintrans$fitted.values + 1)/lamdaopt)
dev.new()
plot(linreg$fitted.values,invtrans,xlab="Predicted, Untransformed",ylab="Predicted, Inverse
Transformed")

OUTPUT

Call:
lm(formula = Aggtrans ~ D.Curve + D.Bent + S.Curve + S.Bent)
```

FIGURE 4.11
Regression with Box–Cox transformed response variable.   *(Continued)*

```
Residuals:
    Min       1Q    Median       3Q       Max
-0.75165  -0.28703  0.00902  0.26233  0.73390

Coefficients:
              Estimate   Std. Error    t value    Pr(>|t|)
(Intercept)   -0.64975      0.26567     -2.446    0.020541  *
D.Curve        1.35105      0.33584      4.023    0.000359  ***
D.Bent         0.18832      0.48754      0.386    0.702030
S.Curve        0.21535      0.31431      0.685    0.498513
S.Bent         0.06555      0.54341      0.121    0.904783
---
Signif. codes:  0 '***' 0.001 '**' 0.01 '*' 0.05 '.' 0.1 ' ' 1

Residual standard error: 0.4033 on 30 degrees of freedom
Multiple R-squared:  0.4226,     Adjusted R-squared:  0.3456
F-statistic: 5.489 on 4 and 30 DF,  p-value: 0.001941
```

FIGURE 4.11 (CONTINUED)
Regression with Box–Cox transformed response variable.

interactions, improves the predictions made by the model. Figure 4.13 shows R code for the squirrel aggressiveness data with several two-factor interaction cross-products included in the model. The Output of the code follows. Notice that the interaction term is specified by putting a ":" between two regressors, for example, D.Curve:S.Curve. Furthermore, since the *p*-values for all the interaction terms included appear to be larger than 0.05, it is unlikely that their addition makes the model a better predictive tool.

Figure 4.14 is an illustration of an Eastern gray squirrel.

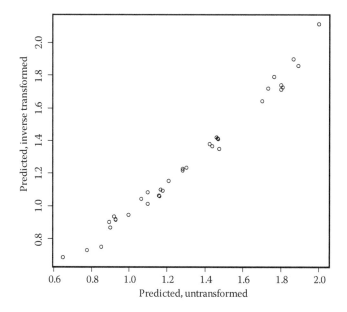

FIGURE 4.12
Inverse-transformed predicted values versus untransformed predicted values.

```
setwd("C:\\Users\\Statistical Methods for Animal Behavior Field Studies\\Statistical Data &
Programs")
df1 <-read.csv("20161005 Example 4.1 Squirrel Aggression Dominant.csv")
#
#
# VARIABLES:
# Aggression
# D.Curve
# D.Bent
# S.Curve
# S.Bent
#
attach(df1)

linreg <-lm(Aggression ~ D.Curve + D.Bent + S.Curve + S.Bent + D.Curve:S.Curve +
D.Curve:S.Bent + D.Bent:S.Curve + D.Bent:S.Bent)
#

OUTPUT

Call:
lm(formula = Aggression ~ D.Curve + D.Bent + S.Curve + S.Bent +
    D.Curve:S.Curve + D.Curve:S.Bent + D.Bent:S.Curve + D.Bent:S.Bent)

Residuals:
    Min       1Q    Median      3Q       Max
-0.26919 -0.04386  0.01655  0.05797   0.15804

Coefficients:
                  Estimate  Std. Error  t value  Pr(>|t|)
(Intercept)       -0.04728    0.13477    -0.351   0.72856
D.Curve            1.92598    0.33057     5.826   3.86e-06 ***
D.Bent             0.90612    0.31651     2.863   0.00819 **
S.Curve           -0.01847    0.25844    -0.071   0.94358
S.Bent             0.65583    0.36083     1.818   0.08067 .
D.Curve:S.Curve    0.23025    0.45600     0.505   0.61787
D.Curve:S.Bent     0.54254    0.91377     0.594   0.55782
D.Bent:S.Curve    -0.12447    0.55051    -0.226   0.82289
D.Bent:S.Bent     -0.75775    0.92317    -0.821   0.41922
---
Signif. codes:  0 '***' 0.001 '**' 0.01 '*' 0.05 '.' 0.1 ' ' 1

Residual standard error: 0.1008 on 26 degrees of freedom
Multiple R-squared:  0.9768,    Adjusted R-squared:  0.9697
F-statistic:  137 on 8 and 26 DF,  p-value: < 2.2e-16
```

FIGURE 4.13
Linear model with interaction terms.

Theoretical Aspects

The linear nature of the model we are using to relate Y to the Z_j can be expressed in matrix/vector notation. Let

$$y = \begin{bmatrix} y_1 \\ \vdots \\ y_n \end{bmatrix}$$

$$\beta = \begin{bmatrix} \beta_0 \\ \beta_1 \\ \vdots \\ \beta_k \end{bmatrix}$$

FIGURE 4.14
Eastern gray squirrel (*Sciurus carolinensis*).

$$Z = \begin{bmatrix} 1 & z_{11} & \cdots & z_{1k} \\ 1 & z_{21} & \cdots & z_{2k} \\ \vdots & \vdots & \vdots & \vdots \\ 1 & z_{n1} & \cdots & z_{nk} \end{bmatrix}$$

$$\varepsilon = \begin{bmatrix} \varepsilon_1 \\ \vdots \\ \varepsilon_n \end{bmatrix}$$

Then the multiple linear regression model can be expressed as a matrix equation:

$$y = Z\beta + \varepsilon$$

The components of the vector β are the unknown coefficients, and the components of the vector ε are the random noise components, indicating that for an individual, even with all the exact same values of the Z_{ij}, the observed value of Y could differ between two or more observations. The least squares estimates of β are given by the equation:

$$\hat{\beta} = [Z^T Z]^{-1} Z^T y$$

$\hat{\beta}$ is called the ordinary least square (OLS) estimator. The *T* indicates matrix transpose, and –1 the matrix inverse. The predicted value of Y for a given set of values for the Z_j, call them $z_1, z_2, \ldots, z_k$, would be

$$\hat{y} = \hat{\beta}_0 + \hat{\beta}_1 z_1 + \hat{\beta}_2 z_2 + \ldots + \hat{\beta}_k z_k$$

Inasmuch as Y is a random variable, the estimates of the components of β are also random variables. As such, confidence intervals for the parameters can be constructed. In order to construct such intervals, an assumption must be made about the random components, ε_i, namely that these noise variables are all independent of each other (the noise in the measurement for one individual is not affected by the noise in another), and that in fact they all are normally distributed with mean $\mu = 0$, and they all have exactly the same standard deviation, σ. Another way of saying this is that the random variables ε_i are independent and identically distributed, or i.i.d. for short, as $N(0, \sigma)$ (a short-hand for Normal distribution with mean 0 and standard deviation σ). The effect of this i.i.d. normality is that the magnitude of variability is about the same for every individual's measurement Y, and that the variation is just as likely to give you a higher than the individual's average value of Y or lower than average value for any given individual.
    If

$$z = \begin{bmatrix} 1 \\ z_1 \\ \vdots \\ z_k \end{bmatrix}$$

represents a particular set of values for the Zj, then the predicted value for Y can be expressed as

$$\hat{y} = \hat{\beta}^T z$$

The variance of the predicted value, given z, is

$$V(\hat{y}) = \sigma^2 \left(z^T (Z^T Z)^{-1} z \right)$$

The value of σ is estimated using computer programs such as the R function *lm()*. The variance/covariance matrix of the model parameter estimates is given by

$$V[\hat{\boldsymbol{\beta}}] = [\mathbf{Z}^T\mathbf{Z}]^{-1}\sigma^2$$

So, the estimated variance/covariance matrix is

$$\hat{V}[\hat{\boldsymbol{\beta}}] = [\mathbf{Z}^T\mathbf{Z}]^{-1}\hat{\sigma}^2$$

The estimated standard error for $\hat{\beta}_j$ is the square root of the jth diagonal element of the matrix $\hat{V}[\hat{\boldsymbol{\beta}}]$. Call this standard error $se(\hat{\beta}_j)$. Then we can test the hypothesis

$$H_0:\ \beta_j = 0$$

against the alternative

$$H_1:\ \beta_j \neq 0$$

by computing the statistic

$$t = \frac{\hat{\beta}_j - 0}{se(\hat{\beta}_j)}$$

We compute a p-value for this statistic using a (central) Student's t distribution with $n - (k + 1)$ degrees of freedom (n = sample size = number of response values, k = number of parameters in model, other than the intercept term). The 100(1–α)% confidence interval for the predicted value of Y given $\mathbf{z}$ is

$$\hat{y} \pm t_\nu\hat{\sigma}\sqrt{\left(\mathbf{z}^T(\mathbf{Z}^T\mathbf{Z})^{-1}\mathbf{z}\right)}$$

The symbol t_ν is the 100(1 – α/2) percentile of a Student's t distribution with $\nu = n - (k + 1)$ degrees of freedom.

Generally there is no physical law or rule that dictates which regressor variables actually can be used to effectively predict the response (even on the average). So, when a collection of regressors is chosen, and the parameters are estimated, the researcher must somehow determine if the selected regressors are meaningful with respect to predicting the response.

There are several tests. First, we must test the assumption that the magnitude of noise is not different for different combinations of regressor values. To do this, first compute all the predicted values of the response, using the estimated parameter values, and compare them to the associated actual response values. So, if y_i represents the observed value of the response for individual i, and $\hat{y}_i$ is the predicted value for that individual, then the residual for individual i is the difference:

$$e_i = y_i - \hat{y}_i$$

A plot of the e_i against the $\hat{y}_i$ will tell much about the assumption of constant noise variance. If the plot shows that the e_i seem to be just as likely to be negative as positive, and that the dispersion of the e_i seems to be fairly across all the predicted values $\hat{y}_i$, then we would feel fairly confident about the assumptions that the noise is distributed with mean 0 and the standard deviation of noise is a constant.

Multiple R-squared (R^2) and adjusted R-squared $\left(R_a^2\right)$ are measures of linearity, that is, measures of how strongly to believe that the response variable (y) can be expressed as a linear combination of the regressors (z_j). The total sums of squares (SST) for all the response observations, "corrected" for the intercept, is

$$SST = \sum_{i=1}^{n}(y_i - \bar{y})^2 = y^T y - n\bar{y}^2$$

The sums of squares for the model, again "correcting" for the intercept, can be expressed as

$$SSR = \hat{\beta}Z^T y - n\bar{y}^2$$

The residual sums of squares is the difference:

$$RSS = SST - SSR = y^T y - \hat{\beta}Z^T y$$

It turns out that least squares is the method of finding the values of the coefficients that, assuming the linear model is correct, minimizes RSS.

A measure of the goodness of the linearity hypothesis is

$$R^2 = \frac{SSR}{SST} = 1 - \frac{RSS}{SST}$$

The idea is that if the linear model is correct, then RSS should be small relative to SST. Clearly, since both RSS and SST are sums of squared terms (why?), R^2 should never be negative. In fact, the smallest RSS could possibly be is 0, so that the largest R^2 could be is 1. So, $R^2 = 1$ implies a perfect linear, and $R^2 = 0$ implies absolutely no linear fit (i.e., no linear relationship between the response and the regressors). The adjusted R-squared is R^2 "adjusted" for sample size and numbers of parameters. If there are k regressors, the total number of parameters, including the intercept, is $r = k + 1$. The degrees of freedom for the total data set is $n - 1$. The degrees of freedom for the model is $k = r - 1$ (total number of parameters minus 1). The degrees of freedom for the residuals is the difference, $n - 1 - (r - 1) = n - r$. The adjusted R-squared is

$$R_a^2 = 1 - \frac{RSS/n - r}{SST/n - 1}$$

In general, $R^2 < R_a^2$. Neither of these measures is the definitive adjudicator of linearity. They are, however, relatively easy to compute, and come along with the summary output from the *lm* function of R. In general, we want both these measures to be closer to 1 than to 0. However, the values of both R^2 and R_a^2 will increase as more and more regressors are included in a model. Inclusion of so many regressors so that R^2 is artificially increased is called *overparameterization*. Overparameterized models will have small residuals (differences between predicted and observed response values) but tend to be poor predictors of future response values (Judge et al., 1985). Using the smallest number of regressors or factors that allow for a "sufficient" fit is called the *principle of parsimony*. In the famous words of William Ockham (circa 1350s): "Pluralitas non est ponenda sine necessitate," or in other words, don't add complexity if you don't need it.

The big question is, what constitutes "sufficient"? Unfortunately, there is no general answer. Perhaps the best a model-builder can do is to decide a priori on the average, how big can residuals be and still be small enough? Then, choose regressors in such a fashion so that the average residulas are not significantly different from the desired value. Of course, while tradition-ally p-values less than 0.05 are considered significant, the level of significance is itself a subjective determination.

A concept related to R^2 is correlation. The correlation between two random variables, Y and V is

$$Corr(Y, W) = \frac{E\left((Y - \mu_Y)(W - \mu_W)\right)}{\sqrt{\sigma_Y^2 \sigma_W^2}}$$

The sample estimate of correlation is

$$\hat{C}(Y,W) = \frac{\frac{1}{n}\sum_{i=1}^{n}(y_i - \bar{y})(w_i - \bar{w})}{\sqrt{s_Y^2 s_W^2}}$$

Key Points for Chapter 4

- A linear model is linear in terms of parameters; the right-hand-side (RHS) variables can be nonlinear functions of other variables.
- RHS variables are called regressors; the single left-hand side variable is called the response.
- Regressors cannot be linear functions of other regressors in the same model; such relationships induce something called collinearity, which precludes the use of least squares.
- Least squares is a common method for obtaining estimates of model parameters using data.
- By making assumptions about the distribution of residual noise, confidence intervals can be computed for the parameters and for predicted values of the response.
- Models, once estimated, should be checked for goodness of fit; a very common measure, albeit not really sufficient, is adjusted R^2.

Exercises and Questions

1. Sometimes it is desirable to "normalize" regressors so that they all fall on the same scale. If M is the median value for regressor X, and the range of values for X, going from smallest (X_{min}) to largest (X_{max}) is $R = X_{max} - X_{min}$, one such normalizing transformation is called the *Helmert transformation* (Pardo and Pardo, 2016):

$$Z = \frac{X - M}{\frac{1}{2}R}$$

Perform the Helmert transformation on the Example 4.1 regressors and fit the model. Do the conclusions change?

2. When regressors are transformed to have the same scale, the magnitude of the coefficients in the regression indicate the importance/influence of each regressor. In the transformed squirrel data, rank the regressors in terms of their influence on the response.

5

The Linear Model:
Discrete Regressor Variables

General Ideas

Often in designed experiments, a set of discretely identified conditions or stimuli are presented to individuals, and a response is measured for each individual. In the case of two stimuli (e.g., an experimental and control condition) a t-test can be used to decide if, on the average, there is a difference in the response variable depending on the stimulus. There are two possible scenarios for which the t-test may be inadequate:

1. There are more than two stimuli to be compared.
2. There is more than one condition, or more than one type of stimulus, that is being applied simultaneously.

The case of scenario (1) is fairly intuitive; for example, one might wish to compare an individual's responses to playbacks of three or more different types of alarm calls. For scenario (2), perhaps in addition to the type of alarm call being played back, the age of the individual who made the call must be considered as well. Animals might respond differently to the same alarm call given by a juvenile as opposed to an adult.

Consider a situation where there is a single "factor" that has $k > 2$ discrete states. The states will be referred to as levels. As in regression, we postulate a linear model to describe how the response variable, y, might be affected by the different levels of the "factor." The model is

$$y_{ij} = \mu + \tau_i + \varepsilon_{ij}$$

The symbol μ represents the "overall" average value of y. The τ_i represents the average amount of difference in y from μ, when the state of the factor (experimental treatment, or condition over which the observations on y are being made), is the ith level. The variable ε represents random fluctuations around the average. The double subscripts on y and ε are needed to represent the fact that for each treatment level i, there are several replicate observations made.

Examples with R Code

When there is a single factor in an experiment with multiple levels, the experiment is referred to as a "one-way" layout. As an example, consider the amount time that the common octopus (*Octopus vulgaris*) takes to open a bivalve prey item. Handling time for bivalve prey varies considerably, and can affect the value of a particular prey item for the octopus (McQuaid 1994). Suppose that octopuses are presented with three different size classes of bivalves: small (20–35 millimeters), medium (36–50 millimeters), and large (51–65 millimeters). The response, y, is the time in seconds that it takes the octopus to open the bivalve (either by prying the two halves of the shell apart or by drilling a hole through the shell). In this example, suppose n = 68 octopuses were observed opening one bivalve from each of the three treatment levels: small, medium, and large. The classical analysis of this kind of experiment is called Analysis of Variance, or ANOVA (Montgomery, 2001). Although it is called analysis of variance, ANOVA is actually a way of deciding whether the means of several groupings of data are the same or not. Like regression, the formulas are derived via the process of least squares, although it is not necessarily obvious. The concept of ANOVA can be summarized as partitioning the total sums of squared differences from the overall average into two basic terms:

1. The sums of squared differences between group averages and the overall average.
2. The sums of squared differences between each individual response result and its corresponding group average.

The R function *anova*() will compute the appropriate p-values. First the *lm*() function must be employed (as mentioned earlier, ANOVA is really least squares in disguise). The code in Figure 5.1 shows how the ANOVA can be computed with R.

The Df column in the analysis of variance table is the degrees of freedom. The Sum_Sq column has sums of squares (for Treatment, it is *SSM*; for Residuals, it is *SSE*). The Mean_Sq is called the Mean Square, and is the Sums of squares divided by the associated degrees of freedom. The F_value column is the value of the F statistic, and Pr(>F) is the *p*-value.

So, with a *p*-value $< 2.2 * 10^{-16}$, we reject the null hypothesis. That is nice to know, but not really enough. All we know is that there is probably some

```
setwd("C:\\Users\\Statistical Methods for Animal Behavior Field Studies\\Statistical Data &
Programs")
df1 <-read.csv("20170910 Example 5.1Handling Time One Way Octopi.csv")
#
#
# VARIABLES:
# Treatment
# Time
#
attach(df1)

linaov <-lm(Time ~ Treatment)

anova(linaov)


OUTPUT

Analysis of Variance Table

Response: Time
            Df    Sum Sq    Mean Sq    F value       Pr(>F)
Treatment    2     78478      39239     373.06     < 2.2e-16 ***
Residuals  201     21142        105
---
Signif. codes:  0 '***' 0.001 '**' 0.01 '*' 0.05 '.' 0.1 ' ' 1
```

FIGURE 5.1
One-way ANOVA in R.

difference between at least two of the three groupings or treatments. We would like to know in particular which pair of treatments differ from each other in terms of the response. If that is so, why not just do t tests for all possible combinations of two treatments? The problem is called the *multiple comparison problem*. The more comparisons you make, the more likely it is that you will have some p-values below the threshold of significance (usually 0.05) even if the null hypothesis is true. There are many ways to compensate for the multiple comparison problem. One possible method is based on a theorem called the *Bonferroni inequality* (Law and Kelton, 2015). If there are k comparisons to make, and it is desired to have a significance level of α over all the comparisons combined, then the Bonferroni inequality says the significance level for each of the k comparisons should be α/k. This Bonferroni adjustment to α is very conservative, meaning that it makes determing significance harder than it really ought to be. A more reasonable method is called *Tukey's honestly significant difference*, or HSD. The R function, *TukeyHSD()*, implements the method. It requires that rather than using the *lm()* and *anova()* functions, you must use the *aov()* function to perform the ANOVA calculations. Figure 5.2 shows the R code and associated output with the one-way data.

The ANOVA table is as it was before. The *TukeyHSD()* function provides the difference in average response between all pairs of treatment levels, together with confidence intervals for the differences (lwr, upr are the limits) and p-values for the differences. The confidence intervals have an overall confidence level of 95 percent, as opposed to each interval having a 95-percent

```
setwd("C:\\Users\\Statistical Methods for Animal Behavior Field Studies\\Statistical Data &
Programs")
df1 <-read.csv("20170910 Example 5.1 Extraction Time One Way Octopi.csv")
#
#
# VARIABLES:
# Treatment
# Time
#
attach(df1)


linaov <-aov(Time ~ Treatment)

anova(linaov)

TukeyHSD(x=linaov,which="Treatment")
plot(TukeyHSD(x=linaov,which="Treatment"))


OUTPUT
Analysis of Variance Table

Response: Time
              Df     Sum Sq    Mean Sq       F value        Pr(>F)
Treatment      2      78478      39239        373.06     < 2.2e-16 ***
Residuals    201      21142        105
---
Signif. codes:  0 '***' 0.001 '**' 0.01 '*' 0.05 '.' 0.1 ' ' 1
>
> TukeyHSD(x=linaov,which="Treatment")
  Tukey multiple comparisons of means
    95% family-wise confidence level

Fit: aov(formula = Time ~ Treatment)

$Treatment
              diff         lwr          upr       p adj
LRG-SML    47.89118    44.64953     51.13282        0
MED-SML    20.63382    17.39218     23.87547        0
MED-LRG   -27.25735   -30.49900    -24.01571        0
```

FIGURE 5.2
One-way ANOVA in R with Tukey's HSD.

confidence level. This means that each individual interval has more than a 95-percent confidence level. The *p*-values in the "p adj" column are also "adjusted" to compensate for the fact that more than one comparison is being made from the same experimental dataset. The *plot()* function makes a graph of the confidence intervals for the differences in means for each pair of groupings. Figure 5.3 shows the plot.

One of the convenient features of the multiple comparison confidence intervals is that any two that do not overlap each other are significantly different

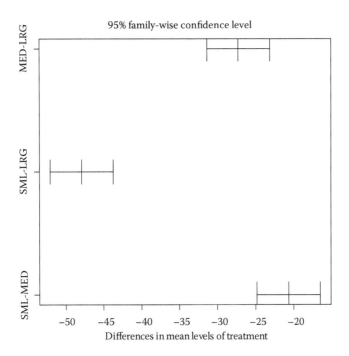

FIGURE 5.3
Tukey HSD confidence interval plot.

from each other at the $100\alpha\%$ level. Thus, in this experiment, the average difference in time between Large and Small treatments is significantly different from the difference between Medium and Small treatments, which in turn is different from the difference between Medium and Large treatments. Furthermore, the difference between Medium and Small treatments is not zero. Since the upper limit for the Medium–Small interval is less than zero, we can conclude that the time spent on opening a bivalve is less when the bivalve is Medium compared to when the bait is Small.

In some cases, the only pairwise comparisons of interest are between most of the treatment groups and a "control" group. For example, suppose the comparison between Medium and Small is not of interest, and the only comparisons of interest are Medium–Large and Small–Large. In that case, Tukey's HSD would be overly conservative. The post-ANOVA pairwise comparison method that is appropriate for comparing all treatment groups to a control is called Dunnett's t (Montgomery, 2001). In the R package called multcomp is a function called *glht()* (general linear hypothesis test). The code and output for the extraction time data are shown in Figure 5.4.

Unlike multiple regression, the object of ANOVA is classification. That is, ANOVA attempts to determine whether or not the groupings induce a repeatable difference in the average value of the response.

```
setwd("C:\\Users\\SMFSBE\\Statistical Data & Programs")
df1 <-read.csv("20170910Example 5.2.b Handling Time One Way Dunnett Octopi.csv")
#
#
# VARIABLES:
# Treatment
# Time
#
#
library(multcomp)

attach(df1)


linaov <-aov(Time ~ Treatment)

anova(linaov)
#
#  The glht function will assume that the factor level with the name
#  having the lowest alphanumeric value is the control.
# In the dataset, the levels of Treatment are therefore named:
# 1LRG
# 2SML
# 3MED
#
testcntl <-glht(linaov,lnfct=mcp(Treatment="Dunnett"))

summary(testcntl)
confint(testcntl)

OUTPUT

> summary(testcntl)

    Simultaneous Tests for General Linear Hypotheses

Fit: aov(formula = Time ~ Treatment)

Linear Hypotheses:
                     Estimate       Std. Error       t value       Pr(>|t|)
(Intercept) == 0      68.674           1.244           55.22       <2e-16 ***
Treatment2SML== 0    -47.891           1.759          -27.23       <2e-16 ***
Treatment3MED== 0    -27.257           1.759          -15.50       <2e-16 ***


---
Signif. codes:  0 '***' 0.001 '**' 0.01 '*' 0.05 '.'     0.1 ' ' 1

(Adjusted p values reported      -- single -step method)

> confint(testcntl)

      Simultaneous Confidence Intervals

Fit: aov(formula = Time ~ Treatment)

Quantile = 2.3283
95% family-wise confidence level

Linear Hypotheses:

                      Estimate         lwr           upr
(Intercept) == 0       68.6735       65.7779       71.5692
Treatment2 SML== 0    -47.8912      -51.9863      -43.7961
Treatment3 MED == 0   -27.2574      -31.3525      -23.1622
```

FIGURE 5.4
One-way ANOVA with Dunnett's t comparisons.

More Than One Treatment: Multiple Factors

Another reason for using ANOVA is the situation where data can be classified not only by a single grouping factor, but by two or more such factors. If there were two such factors, the linear model could be expressed as

$$y_{ijk} = \mu + \tau_i + \gamma_j + \varepsilon_{ijk}$$

The variable y_{ijk} is the kth replicate value of the response for level i of treatment τ and level j of treatment γ. As in the case of the single factor, the τ_i represent the average difference form overall average, μ, when the individual is exposed to level i of factor τ, and γ_j is the average difference from μ when the individual is exposed to level j of factor γ. It is possible that the difference from μ not only depends on "treatments" τ_i and γ_j, but there is also a different effect of τ_i when factor γ is at level j. Such a differential effect of one factor dependent on the level of another is in fact an interaction, and can also be included in the linear model as

$$y_{ijk} = \mu + \tau_i + \gamma_j + \tau\gamma_{ij} + \varepsilon_{ijk}$$

In general, null hypotheses can be simultaneously tested for each of the terms in the model. The null hypotheses would have the form:

$$H_0: \tau_i = 0$$
$$H_0: \gamma_j = 0$$
$$H_0: \tau\gamma_{ij} = 0$$

We will let q = the number of levels of factor τ, and r = the number of levels for factor γ.

To help bring some intuitive focus into the two-factor model, suppose there are two factors, τ and γ, each having $q = r = 2$ levels. Suppose further that the overall (population) mean is $\mu = 100$. Now consider the table of effects, Table 5.1.

The average effect of τ, or change from the overall average μ, is called the main effect of factor τ. Similarly, the average "effect" of γ, or change from the overall average μ, is called the main effect of factor γ.

The expected value of the response, y, depends on the levels of the two factors. For example:

$$E[y_{11}] = \mu + \tau_1 + \tau\gamma_{11} = 100 + -15 + -5 = 80.0$$

However, the expected value of y when $i = 1$, which we will denote $y_{i.}$, averaged over all values of j, is

$$E[y_{1.}] = \mu + \tau_1 + \tau\gamma_{11} + \tau\gamma_{12} = 100 + -15 + -5 + 20 = 100.0$$

TABLE 5.1

Some Hypothetical Effects in a Two-Factor Experiment with Interaction

Interaction Effect	Effect Present Value
τ_1	−15
τ_2	15
γ_1	12.5
γ_2	−12.5
$\tau\gamma_{11}$	−5
$\tau\gamma_{12}$	20
$\tau\gamma_{21}$	5
$\tau\gamma_{22}$	−20
$\mu = 100$	

Conversely, suppose the effects are given in Table 5.2.

The expected values of the response under the same conditions are now:

$$E[y_{11}] = \mu + \tau_1 + \tau\gamma_{11} = 100 + -15 + 0 = 85.0$$

and

$$E[y_{1.}] = \mu + \tau_1 + \tau\gamma_{11} + \tau\gamma_{12} = 100 + -15 + 0 + 0 = 85.0$$

Thus, in the first case, the effect of factor τ depends on the level of factor γ. In the second case, the effect of factor τ is unaffected by the level of factor γ.

TABLE 5.2

Some Hypothetical Effects in a Two-Factor Experiment without Interaction

No Interaction Effect	Effect Value
τ_1	−15
τ_2	15
γ_1	12.5
γ_2	−12.5
$\tau\gamma_{11}$	0
$\tau\gamma_{12}$	0
$\tau\gamma_{21}$	0
$\tau\gamma_{22}$	0
$\mu = 100$	

Consider the handling time data. Suppose that it turned out not only were there three different size classes of bivalve prey (Small, Medium, and Large) but there were also two different species of bivalves: Common cockles (*Cerastoderma edule*) and blue mussels (*Mytilus edulis*). The R code for the two factor analysis is shown in Figure 5.5, together with the R console output.

Average response values by level for a given factor can be obtained using the *tapply*() function. Figure 5.6 shows the output for the average Time values for each prey size class and for each prey Prey Species.

The Prey Size effect still appears to be significant, and in fact all levels differ from the others in terms of the average value of Time. In addition, there does seem to be an effect of Prey Species, and an interaction effect between Prey Size and Prey Species (p-value = 0.0005487). The table made

```
setwd("C:\\Users\\Statistical Methods for Animal Behavior Field Studies\\Statistical Data &
Programs")
df1 <-read.csv("20170910 Example 5.3 Handling Time Two Factors.csv")
#
#
# VARIABLES:
# Treatment
# Prey
# Time
#
attach(df1)


linaov <-aov(Time ~ Treatment + Prey + Treatment:Prey)

anova(linaov)

TukeyHSD(x=linaov,which=c("Treatment","Prey"))
plot(TukeyHSD(x=linaov,which="Treatment"))
dev.new()
plot(TukeyHSD(x=linaov,which="Prey"))

OUTPUT

Analysis of Variance Table

Response: Time
                Df    Sum Sq   Mean Sq    F value      Pr(>F)
Treatment        2     78478     39239   612.4581   < 2e-16 ***
Prey             1      8028      8028   125.2975   < 2e-16 ***
Treatment:Prey   2       428       214     3.3437   0.03731 *
Residuals      198     12686     12686       64
--
Signif. codes:  0 '***' 0.001 '**' 0.01 '*' 0.05 '.'0.1 ' ' 1
>
> TukeyHSD(x=linaov,which=c("Treatment","Prey"))
  Tukey multiple comparisons of means
    95% family-wise confidence level

Fit: aov(formula = Time ~ Treatment + Prey+ Treatment:Prey)

$Treatment
              diff       lwr       upr      p adj
LRG-SML   47.89118  44.64953  51.13282      0
MED-SML   20.63382  17.39218  23.87547      0
MED-LRG  -27.25735 -30.49900 -24.01571      0

$Prey
              diff       lwr       upr      p adj
COC-MUS   12.54608   10.3358  14.75636      0
```

FIGURE 5.5
Two-factor ANOVA with Tukey's HSD.

```
> tapply(Time,INDEX=Treatment,FUN=mean)

    SML      LRG      MED

20.78235 68.67353 41.41618

> tapply(Time,INDEX=Prey,FUN=mean)

    MUS      COC

37.35098   49.89706

> tapply(Time,INDEX=list(Treatment,Prey),FUN=mean)

            MUS       COC

SML     16.55882   25.00588

LRG     61.37059   75.97647

MED     34.12353   48.70882
```

FIGURE 5.6
Average time by treatment and by prey type.

by *tapply()* with INDEX=list(Treatment,Container) shows how regardless of Prey Species, the average Time to open small bivalves was lowest, followed by medium-sized bivalves, and then by large bivalves. However, the difference between cockles and mussels was approximately 7 seconds for the Small treatment, 18 seconds for Large, and 15 seconds for Medium. The fact that the interaction term Prey Size:Prey Species had a *p*-value less than 0.05 lends credence to the belief that the difference in handling time between species does in fact depend on the size class of the prey item. These means can be plotted using the *interaction.plot()* function. The function call is

interaction.plot(x.factor=Treatment,trace.factor=Container,response=Time)

The x.factor parameter is the factor plotted on the horizontal axis, and the trace.factor is the factor used to "split" the data. Figure 5.7 shows the resulting plot. The vertical axis is the response variable. The points on the graph are the average response values for each combination of factor levels.

Blocking Factors

Sometimes there are factors that exist in the experimental situation that are a nuisance, inasmuch as they may affect the response, but are not of any interest. For example, males and females might differ in the intensity of

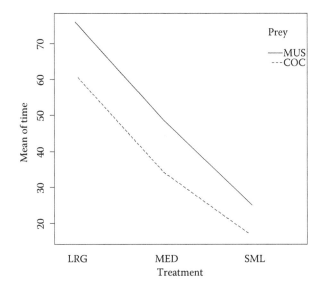

FIGURE 5.7
Interaction plot for treatment and container.

their response to a particular treatment, yet show the same relative differences between treatment factor levels. The nuisance factors will be referred to as blocks. It is altogether possible that within any block level, the treatment factor levels may have all been applied, and applied in more or less the same numbers. Such a situation is referred to as a complete, balanced block.

Suppose that in the octopus example, the individual octopuses happen to come from one of two distinct areas (A and B). In that case, it might be helpful to know if differences exist between the octopus areas in terms of how long it takes them to pull apart bivalves. If such a difference exists, and it is not included in the model for the ANOVA, the residual error sums of squares could be inflated, making otherwise significant factors appear to be not significant. Figure 5.8 shows the R code with the addition of the Area blocking factor added into the model. Usually, blocks are not considered to have the potential to interact with experimental factors. Thus, in the *lm()* call the model includes the Area factor but no interactions between Area and the experimental factors.

In the last *tapply()* call, all three factors, Prey Size, Prey Species, and Octopus Size were listed in the INDEX parameter. Thus the analyst can see that the mean extraction time is higher in all categories for Area B than for Area A. It is typical of "blocking" factors that while the effects of experimental factors are the same, statistically speaking, for every block, the average difference between blocks is different.

```
setwd("C:\\Users\\Statistical Methods for Animal Behavior Field Studies\\Statistical Data &
Programs")
df1 <-read.csv("20161103 Example 5.4 Extraction Time Two Factors with Blocks.csv")
#
#
# VARIABLES:
# AreaArea
# Treatment
# Prey
# Time
#
attach(df1)


linaov <-aov(Time ~ Area + Treatment + Prey + Treatment:Prey)

anova(linaov)

TukeyHSD(x=linaov,which=c("Treatment"," Prey"))

tapply(Time,INDEX=Treatment,FUN=mean)
tapply(Time,INDEX=Prey,FUN=mean)
tapply(Time,INDEX=Area,FUN=mean)
tapply(Time,INDEX=list(Treatment,Prey,Area),FUN=mean)
plot(TukeyHSD(x=linaov,which="Treatment"))
dev.new()
plot(TukeyHSD(x=linaov,which="Prey"))
dev.new()
plot(TukeyHSD(x=linaov,which="Area"))
dev.new()
interaction.plot(x.factor=Treatment,trace.factor=Prey,response=Time)

OUTPUT

Analysis of Variance Table

Response: Time
```

	Df	Sum Sq	Mean Sq	F value	Pr(>F)
Area	1	5158	5158	131.7816	< 2.2e-16 ***
Treatment	2	77966	38983	995.9472	< 2.2e-16 ***
Prey	1	8309	8309	212.2795	< 2.2-16 ***e
Treatment:Prey	2	476	238	6.0813	0.002737 **
Residuals	197	7711	39		

```
---
Signif. codes:  0 '***' 0.001 '**' 0.01 '*' 0.05 '.' 0.1 ' ' 1>

> TukeyHSD(x=linaov,which=c("Treatment","Prey"))
 Tukey multiple comparisons of means
 95% family-wise confidence level

Fit: aov(formula = Time ~ Area+ Treatment + Prey+ Treatment:Prey)


$Treatment
```

	diff	lwr	upr	p adj
LRG-SML	47.89118	44.64953	51.13282	0
MED-SML	20.63382	17.39218	23.87547	0
MED-LRG	-27.25735	-30.49900	-24.01571	0

```
$Prey
```

	diff	lwr	upr	p adj
Tube-Box	12.54608	10.3358	14.75636	0

```
>
> tapply(Time,INDEX=Treatment,FUN=mean)
    SML        LRG        MED
20.78235   68.67353   41.41618
> tapply(Time,INDEX=Prey,FUN=mean)
    MUS        COC
37.35098   49.89706
> tapply(Time,INDEX=Area,FUN=mean)
    LRG        SML
 35.02692   46.56513
```

FIGURE 5.8

Extraction time code with Area-blocking factor and output.   *(Continued)*

```
> tapply(Time,INDEX=list(Treatment,Prey,Area),FUN=mean)
, , LRG

               MUS            COC
SML       12.51111       17.90000
LRG       55.95000       63.52222
MED       27.13750       34.58889

, , SML

               MUS            COC
SML       18.01600        27.564
LRG       63.03846        80.460
MED       36.27308        53.792
```

FIGURE 5.8 (CONTINUED)
Extraction time code with Area-blocking factor and output.

ANOVA and Permutation Tests

As in the case of multiple regression with continuously valued regressors, the *p*-values from ANOVA require the assumptions that the residuals from the linear model are normally distributed and that the standard deviation of the residuals is constant, regardless of the levels of the factors. If these assumptions are violated, those *p*-values may not be valid. As in the case of the two group comparisons using *t*-tests and their variants, a permutation test can be performed with ANOVA, which circumvents the need to meet the above assumptions. Consider the function *aov*(). It creates an aov object, which has attributes. The *anova*() function also creates an object. If we make the following assignment in the code:

aov1 <- anova(linaov)

then the object aov1 has some attributes that can be captured and stored. To find the names of the attributes, we can use the function attributes(), as in Figure 5.9.

In particular, the attribute called "F value" is a vector with F statistics for each term in the model used to create the *aov*() object. Since Prey Size is the second term in the model specification, the variable aov1$"F value" [2] has its associated F statistic. Using the *rand*() and *order*() functions, we can generate a permutation null distribution for the Prey Size F statistic, and obtain a nonparametric *p*-value for Prey Size. Figure 5.10 shows the R code for making the permutation test computations.

The variable p.val has the nonparametric *p*-value for Treatment. Its value is approximately 0, thus confirming the parameter ANOVA *p*-value. The function *hist*(F.null) made the histogram shown in Figure 5.11.

The original F statistic for Prey Size was 1047.592 which is clearly well beyond the 95th percentile of the permutation null distribution. Hence the conclusion is that there is a significant difference in average Handling Time among the three prey size class "treatments."

```
> attributes(aov1)
$names
[1] "Df" "Sum Sq"  "Mean Sq" "F value" "Pr(>F)"

$row.names
[1] "Area"       "Treatment"      "Prey"
[4] "Treatment:Container" "Residuals"

$class
[1] "anova" "data.frame"

$heading
[1] "Analysis of Variance Table\n" "Response: Time"
```

FIGURE 5.9
The "attributes" Function for Object "aov1."

```
setwd("C:\\Users\\Statistical Methods for Animal Behavior Field Studies\\Statistical Data &
Programs")
df1 <-read.csv("20170910 Example 5.4 Handling Time Two Factors with Blocks.csv")
#
#
# VARIABLES:
# Group
# Treatment
# Container
# Time
#
attach(df1)
permute.time <-c()
F.null <c()

linaov <-aov(Time ~ Group + Treatment + Container + Treatment:Container)

aov1 <-anova(linaov)

TukeyHSD(x=linaov,which=c("Treatment","Container"))

F.original <-aov1$"F value"[2] # this is the Unpermuted F value for the Treatment Effect

nsize <-nrow(df1)
#
#  generate permutation distribution
#
nreps <-1000

for (i in1:nreps) {
 rand <-runif(nsize,0,1) #obtains random numbers between 0 and 1
 permute.time <-Time[order(rand)] # this permutes the response data
 df.permute <-data.frame(Group,Treatment,Container,permute.time) #this creates a permuted
dataframe
 lin.perm <-aov(df.permute$permute.time ~ df.permute$Group + df.permute$Treatment +
df.permute$Container + df.permute$Treatment:df.permute$Container)
 aov.perm <-anova(lin.perm)
 F.null[i] <-aov.perm$"F value"[2]
 }
p.val <-sum(F.null >= F.original)
p.val <-p.val / nreps
```

FIGURE 5.10
Extraction time permutation test for treatment.

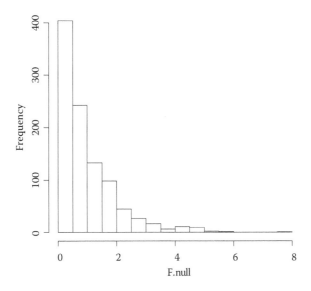

FIGURE 5.11
Histogram of permutation null F statistics for treatment.

Nested Factors

Sometimes the levels of a factor depend on the levels of another. For example, suppose that within each Area, individuals were presented with all three size classes of bivalve prey, and with both species of bivalves. Perhaps the variation is large enough to inflate the noise level, and thus increase the residual sums of squares (SSE) artificially. The individual identifications are said to be nested within Area. That is, the between-individual effect may be different, depending on which Area the individuals come from. Thus, the linear model should account for the between-individual effect within each Area. In this case, suppose there were 17 individuals within each group. The degrees of freedom associated with a nested factor are the total number of levels in the "outer" nesting factor times the number of levels of the nested factor minus one. In the example, there were two Areas, and 17 individuals nested within Octopus Size. Thus, the degrees of freedom for the individuals factor is $2 * (17 - 1) = 32$. The R code for including a nested factor, together with its output, are given in Figure 5.12.

```
setwd("C:\\Users\\Statistical Methods for Animal Behavior Field Studies\\Statistical Data &
Programs")
df1 <-read.csv("20170910 Example 5.6 Handling Time Nested Factor.csv")
#
#
# VARIABLES:
# Area
# Individual (nested within Area)
# Treatment
# Prey
# Time
#
attach(df1)

fIndiv <-factor(Individual) #need to coerce this variable
#               so it will be treated as discrete, not continuous
#
linaov <-aov(Time ~ Area+ fIndiv%in%Area+ Treatment + Prey + Treatment:Prey)

anova(linaov)

TukeyHSD(x=linaov,which=c("Treatment","Prey"))

tapply(Time,INDEX=Treatment,FUN=mean)
tapply(Time,INDEX=Prey,FUN=mean)
tapply(Time,INDEX=Area,FUN=mean)·
tapply(Time,INDEX=list(Treatment,Prey,Area),FUN=mean)
plot(TukeyHSD(x=linaov,which="Treatment"))
dev.new()
plot(TukeyHSD(x=linaov,which="Prey"))
dev.new()
plot(TukeyHSD(x=linaov,which="Area"))
dev.new()
interaction.plot(x.factor=Treatment,trace.factor=Prey,response=Time)


OUTPUT

Analysis of Variance Table

Response: Time
```

	Df	Sum Sq	Mean Sq	F value	Pr(>F)	
Area	1	2903	2903	81.7411	4.072e Area -16	***
Treatment	2	78478	39239	1105.0290	< 2.2e-16	***
Prey	1	8028	8028	226.0683	< 2.2e-16	***
Area:fIndiv	32	3924	123	3.4532	1.092e-07	***
Treatment:Prey	2	428	214	6.0329	0.00296	**
Residuals	165	5859	36			

```
---
Signif. codes:  0 '***' 0.001 '**' 0.01 '*' 0.05 '.' 0.1 ' ' 1
```

FIGURE 5.12
Extraction time with individuals nested in Area.

Analysis of Covariance: Models with Both Discrete and Continuous Regressors

Sometimes when performing an experiment with a discrete, categorical factor, some measureable and continuous quantity may be varying without control. For example, ambient temperature will vary, and in the field is not controllable. Such variables, called *covariates*, may affect the response, but are not of interest. The linear model may incorporate such variables, in order to

```
setwd("C:\\Users\\SMFSBE\\Statistical Data & Programs")
df1 <-read.csv("20170910 Example 5.7 Handling Time One Way with Covariate.csv")
#library(car)
#
#
# VARIABLES:
# Individual
# Treatment
# Temp
# Time
#
attach(df1)


onewayaov <-lm(Time ~ Treatment)
ancova.model <-lm(Time ~ Treatment + Temp)

Anova(onewayaov,type="III")
Anova(ancova.model,type="III")
TimeSMLSML<-Time[Treatment=="SMLSML"]
TempSMLSML<-Temp[Treatment=="SMLSML"]
TimeMED<-Time[Treatment=="MED"]
TempMED<-Temp[Treatment=="MED"]
TimeLRG<-Time[Treatment=="LRG"]
TempLRG<-Temp[Treatment=="LRG"]


plot(TempSMLSML,TimeSMLSML,pch=1,main="One Way ANCOVA
plot",xlim=c(25,32),ylim=c(min(Time)-1,max(Time)+1),xlab="Temp (oC)",ylab="Time (sec)")
points(TempMED,TimeMED,pch=2)
points(TempLRG,TimeLRG,pch=3)

OUTPUT

>
>
> onewayaov <-lm(Time ~ Treatment)
> ancova.model <-lm(Time ~ Treatment + Temp)
>
> Anova(onewayaov,type="III")
Anova Table (Type III tests)

Response: Time
              Sum Sq     Df    F value     Pr(>F)
(Intercept)   103842      1   32377.4680   <2e-16 ***
Treatment         14      2      2.1276    0.1246
Residuals        311     97
---
Signif. codes:  0 '***' 0.001 '**' 0.01 '*' 0.05 '.' 0.1 ' ' 1
> Anova(ancova.model,type="III")
Anova Table (Type III tests)

Response: Time
               Sum Sq    Df    F value     Pr(>F)
(Intercept)   173.789     1    76.8724    6.584e-14 ***
Treatment      18.763     2     4.1498    0.01868 *
Temp           94.071     1    41.6103    4.503e-09 ***
Residuals     217.032    96
---
Signif. codes:  0 '***' 0.001 '**' 0.01 '*' 0.05 '.' 0.1 ' ' 1
```

FIGURE 5.13
Analysis of covariance example.

"correct" for their effects. Such analyses are referred to as analysis of covariance (ANCOVA). Suppose the octopus example had a single factor, Prey Size (Small, Medium, Large). However, also suppose that temperature had some effect on prey handling time. The code in Figure 5.13 shows the analyses with just the Prey Size effect and with the Prey Size + Temp effect, plus the output from the *Anova()* function in library "car."

Without the Temp covariate, the *p*-value for Treatment was 0.1246. When the covariate was included, the *p*-value became 0.01868. Figure 5.14 shows a

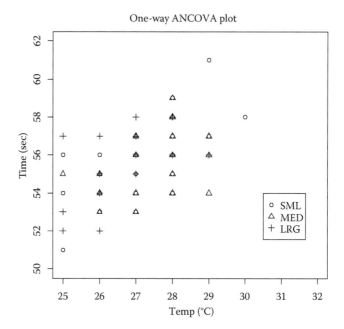

FIGURE 5.14
Time versus temperature: ANCOVA example.

plot of the response (Time), ploted against temperature (Temp), for each of the three treatments.

Although the plot does not necessarily make the differences between the treatments obvious, it does seem clear that the response is varying with temperature. In general, covariates are included in models to avoid masking significant effects of discrete factors.

Theoretical Aspects

Multiple Groupings: One-Way ANOVA

If y_{ij} is the jth observation of the response in group i, $\overline{\overline{Y}}$ is the average of all the data, regardless of grouping, and $\bar{y}_i$ represents the average for group i (in our example, $i = 1, 3$), then

$$SST = \sum_{i=1}^{k} \sum_{j=1}^{n_i} \left(y_{ij} - \overline{\overline{Y}}\right)^2$$

$$SSM = \sum_{i=1}^{k} \left(\bar{y}_i - \bar{\bar{Y}} \right)^2$$

$$SSE = \sum_{i=1}^{k} \sum_{j=1}^{n_i} \left(y_{ij} - \bar{y}_i \right)^2$$

where k = number of groupings (= 3 in this example) and n_i = number or replicate observations in group i (in the example n_i = 68 for all values of i).

SST is called the "total" sums of squares; *SSM* is the "model" sums of squares, and *SSE* the within, residual, or "error" sums of squares. It turns out that

$$SST = SSM + SSE$$

The null hypothesis for the one-way ANOVA is

$$H_0: \tau_i = 0$$

for all $i = 1, k$, so that if it is true, then

$$y_{ij} = \mu + \varepsilon_{ij}$$

regardless of the grouping. If this null hypothesis is in fact true, then SSM should be small compared to SSE. Let $N = \sum_{i=1}^{k} n_i$ be the total sample size. It also turns out that, if the null hypothesis is true, the variables ε_{ij} are normally distributed with expected value of zero, and the variables ε_{ij} all have the same standard deviation (σ), then the ratio

$$F = \frac{SSM/k - 1}{SSE/N - k}$$

has an F distribution with $k - 1$ degrees of freedom for the numerator and $N - k$ degrees of freedom for the denominator. Thus, a p-value can be computed to determine whether to believe the null hypothesis. The sums of squares divided by the associated degrees of freedom is called the *mean square*. For discrete factors, the degrees of freedom for the factor is the number of levels minus one. The total degrees of freedom for the model is $N - 1$, that is, the total sample size minus one. The degrees of freedom for error (residuals) is the total degrees of freedom minus the sum of the degrees of freedom for all the nonerror terms in the model.

Key Points for Chapter 5

- A linear model can be used to describe the relationship between a continuous response variable and discrete regressors.
- Discrete regressors are generally called factors.
- The analysis is usually classificatory and not predictive; the objective is to determine which of the factors actually are associated with differences in average response values.
- The usual method for analysis is called analysis of variance, or ANOVA.
- If a factor is found to be significant, then a more specific test can be performed to determine which values of the factor (called levels) are associated with difference average response values; confidence intervals for the average response at each level of each factor can be constructed.
- Permutation tests can be constructed to compute p-values.
- ANCOVA can be used to compensate for uncontrolled continuous regressors that may mask the signifcant effects of discrete factors.

Exercises and Questions

1. Consider the dataset 20161120 Exercise 5.1 Batesian Mimicry.csv, which is based on a study by Golding and Edmunds (2000). This study investigated the degree to which droneflies (*Eristalis spp.*) mimic the behavior of honeybees (*Apis mellifera*). Droneflies are harmless but look similar to honeybees (which sting), and the authors hypothesized that droneflies have also evolved to mimic the behavior of honeybees in order to fool predators. They measured the amount of time spent on flowers by droneflies, honeybees, and several other species of Dipterans (flies) and Hymenopterans (bees, wasps, and ants). Perform an ANOVA and use Tukey's HSD to determine if there is any difference in the amount of time spent on flowers between droneflies and honeybees. Make an interaction plot for the insect type by flower type.

6

The Linear Model: Random Effects and Mixed Models

General Ideas

So far, our discussion of linear models, continuous regressors, experimental factors, and blocking factors have all been based on the assumption that there is nothing random about right-hand side variables. That is, in the examples of Chapter 4, if a squirrel's tail curvature was measured at 43 percent, it was exactly 43 percent; there was no appreciable error associated with the measurement. Similarly, in Chapter 5, all bivalves in the Small size class were assumed to be effectively the same, without variation. We also implicitly assumed that there were exactly two areas from which the octopuses come, and that these two areas were the only ones under consideration. These kinds of factors or regressors that do not have any randomness associated with them are known as *fixed effects*. Conversely, factors or regressors that do have some randomness are called *random effects*. Including random effects in a model contributes some complexities to the process of predicting the response variable values. Mixed models are models that contain both fixed and random effects.

The linear mixed model looks very much like the multiple linear regression or ANOVA model:

$$y_{ijklm} = \beta_0 + \sum_{i=1}^{q} \beta_i z_i + \sum_{i=1}^{r} \gamma_j x_j + \sum_{k=1}^{s} \tau_k + \sum_{l=1}^{t} \theta_l + \varepsilon_{ijklm}$$

The z_i are continuous, fixed effect regressors, such as those described in the squirrel example of Chapter 4. The x_j are continuously valued random variables whose observed values could vary in a probabilistic fashion even within the same individual. The τ_k are discrete fixed effects such as those described in the octopus example of Chapter 5. The θ_l are discrete random variables that also can vary in a probabilistic fashion, at least from experiment to experiment. The objective is to obtain a predicted value for the response y for given values of the z_i. In the case of fixed effect models, the predicted values are obtained via least squares estimates of the regressor coefficients. Such "predicted" values are referred to as best linear unbiased estimators, or *BLUEs*. When some of the predictors are in fact random variables themselves, the predictions for the response are a little more complicated. They are refererred to as best linear unbiased predictors, or *BLUPs*. The formulas for *BLUEs* and *BLUPs* are different, although in many cases the actual predictions do not differ substantially.

Simple Case: One Fixed and One Random Effect

Like the ANOVA situation described in Chapter 5, suppose the model for the response is

$$y_{ijk} = \mu + \tau_i + \gamma_j + \varepsilon_{ijk}$$

This is a two-factor model without interaction. However, suppose that the effect of one of the factors, say γ, is a random variable, in that levels of the factor were randomly selected. The question is whether this model is analyzed in the same way as when γ was a "fixed" effect; that is, when the levels were treated as though they were in fact the only possible levels for the factor. The answer is no. The idea is that while a fixed effect, τ, contributes to the average value of the response, a random effect contributes to the variability of the response. Consider the following example based on a study by Romano et al. (2016).

Examples with R Code

For many species in which parents provide their dependent offspring with food, parents have limited resources with which to provision their young. If some offspring have a better chance of surviving and reproducing in their own right, parents are expected to invest more heavily in these more "valuable" offspring. In barn swallows (*Hirundo rustica*), males with darker undersides have greater reproductive success. However, for females, underside color is less strongly correlated with reproductive success. Plumage color develops while the birds are still nestlings, and thus serves as a reliable indicator of a male, but not a female, chick's future reproductive value. Romano et al. hypothesized that barn swallow parents would preferentially feed darker male nestlings over lighter-colored male nestlings, but would not discriminate between females based on color. The researchers temporarily removed all chicks except for two same-sex siblings from a sample of 36 nests, such that they had 18 female–female dyads (sisters) and 18 male–male dyads (brothers). They manipulated the color of nestlings' undersides using markers, so that each dyad had one artificially darkened chick and one sham-colored chick (painted with a clear marker). Then they measured the weight of each nestling before and after a 90-minute period in which the parents were allowed to feed the chicks. They also measured the proportion of food items that were delivered to the darkened versus the sham-colored chick.

To simplify the experiment, we will initially ignore sex. Each dyad represents a randomly selected pair of nestlings. The model will only have two factors, the treatment (dark or light undersides) and dyad (1 to 36). The response variable will be weight gain. The dyad factor will be considered as a random effect because the dyads included in the experiment are only a random sample of all the potential dyads of interest (i.e., not every possible pair of same-sex barn swallow siblings was included in the experiment). If we were going to treat dyad as a fixed effect (acting as if the 36 dyads in this experiment were the only ones in existence), then we would use the function *lm()* that was introduced in Chapters 4 and 5. However, because dyad is to be treated as a random effect, will will instead use the the function *lme()*. As a random effect, dyad will be considered to constitute to the overall variation in weight gain. The code in Figure 6.1 and associated output shows how dyad can be treated as a fixed effect and as a random effect.

Two models were fit to the data. The first (mixedf) is a pure fixed-effects model. The two factors are Treatment and Dyad (coerced as a factor, since its levels are coded as numbers 1 through 36; if it were not coerced as a factor, R would treat it as a continuously valued regressor). The second model has a fixed effect (Treatment) and a random effect, Dyad (fDyad). The fixed effect model is the familiar case described in Chapter 5. Both factors have p-values

```
setwd("C:\\Users\\Statistical Methods for Animal Behavior Field Studies\\Statistical Data &
Programs")
df1 <-read.csv("20161114 Example 6.1 Swallow Nestlings Dyad Treatment WG.csv")
attach(df1)
fDyad <-factor(Dyad)

tapply(WG, INDEX=Treatment, FUN=mean)

mixedf <-lm(WG ~ Treatment + fDyad, data = df1)

mixedr <-lme(WG ~ Treatment, data = df1, random = ~1|fDyad)

anova(mixedf)

anova(mixedr)

summary(mixedr)
```

FIGURE 6.1
Swallow nestling model: Treatment and dyad only. *(Continued)*

```

> tapply(WG,INDEX=Treatment,FUN=mean)
    Dark     Light
0.9075000 0.6777778
>
> mixedf <-lm(WG ~ Treatment + fDyad, data = df1)
>
> mixedr <-lme(WG ~ Treatment, data = df1, random = ~1|fDyad)
>
> anova(mixedf)
Analysis of Variance Table

Response: WG
            Df    Sum Sq    Mean Sq    F value    Pr(>F)
Treatment    1    0.9499    0.94990    17.2026    0.0002033 ***
fDyad       35    3.8600    0.11029     1.9973    0.0221001 *
Residuals   35    1.9326    0.05522
---
Signif. codes:  0 '***' 0.001 '**' 0.01 '*' 0.05 '.' 0.1 ' ' 1
>
> anova(mixedr)
              numDF    denDF    F-value    p-value
(Intercept)       1       35    410.1649    <.0001
Treatment         1       35     17.2026    2e-04
>
> summary(mixedr)
Linear mixed-effects model fit by REML
 Data: df1
    AIC       BIC     logLik
 35.27906  44.27304  -13.63953

Random effects:
 Formula: ~1 | fDyad
         (Intercept)  Residual
StdDev:    0.1659346  0.2349862

Fixed effects: WG ~ Treatment
                  Value    Std.Error    DF    t-value    p-value
(Intercept)    0.9075000   0.04794465    35   18.92808    0e+00
TreatmentLight -0.2297222  0.05538679    35   -4.14760    2e-04
 Correlation:
             (Intr)
TreatmentLight-0.578

Standardized Within-Group Residuals:

     Min         Q1        Med        Q3        Max
-1.7732003 -0.6648534 -0.0758223 0.6445873  1.9203134

Number of Observations: 72
Number of Groups: 36
```

FIGURE 6.1 (CONTINUED)
Swallow nestling model: Treatment and dyad only.

less than 0.05. Thus, we would conclude that they both significantly affect the average value of WG (weight gain, in grams). In the second model (mixedr), Dyad is treated as a random effect. The fixed effect, Treatment, is still significant (no change to its *p*-value). The Dyad effect is considered to contribute to the variability in the response (WG). There are several methods that can be used to estimate what proportion of the total variance of the response is due to each of the random effects in the model. Possibly the most popular method is called restricted maximum likelihood (REML). The details of how REML works are described in Searle et al. (1992). The basic idea of REML is that it first "subtracts out" the fixed effects before attempting to estimate the components of variance due to each random effect in the model. In this model, there are two random effects: Dyad and model errors, or residuals. The part of the total variance attributed to Dyad is $\hat{\sigma}_D^2 = 0.1659346^2 \approx 0.02753429$. The residual variance is estimated to be $\hat{\sigma}_\varepsilon^2 = 0.2349862^2 \approx 0.05521851$. The total variance estimate is therefore:

$$\hat{\sigma}_T^2 = \hat{\sigma}_D^2 + \hat{\sigma}_\varepsilon^2$$

One can then compute the percentage of the total variance attributed to each random effect:

$$\frac{\hat{\sigma}_D^2}{\hat{\sigma}_D^2 + \hat{\sigma}_\varepsilon^2} = \frac{0.02753429}{0.02753429 + 0.05521851} \approx 0.3327173 \approx 33.27\%$$

Similarly:

$$\frac{\hat{\sigma}_\varepsilon^2}{\hat{\sigma}_D^2 + \hat{\sigma}_\varepsilon^2} = \frac{0.05521851}{0.02753429 + 0.05521851} \approx 0.6672706 \approx 66.73\%$$

More Complex Case: Multiple Fixed and Random Effects

In the swallow nestling example, the Treatment (light or dark coloring) was a fixed effect and dyad was a random effect. Now consider the addition of four fixed effects: Nestling sex (Dyads had either two female or two male chicks) and three colormetric measurements, called rA, Theta, and Upsilon, which are continuously valued functions of the light reflectance from the underside of the chick. Also, assume that more than one Dyad was selected from each of several nests, which in turn were selected at random. Thus, both Dyad and Nest will be treated as random effects. A more complex model for weight gain could be conceptually expressed as

$$WG_{ijklm} = Treatment_i + Sex_j + Nest_k + Dyad_l(Nest_k)$$
$$+ rA_{ijklm} + Theta_{ijklm} + Upsilon_{ijklm} + \varepsilon_{ijklm}$$

The Dyad effect is "nested" under the Nest effect, since each Nest had its own unique Dyads. The R function *lmer()*, a more general mixed model function than *lme()*, will be used to fit the mixed effects model. The code is given in Figure 6.2. The package lme4 must first be attached via a library(lme4) command. Of course, it also means that in order to use function *lmer*, the package lme4 must be installed.

Notice that in addition to the mixed effects model, a purely fixed effects model was fit using the function *lm()*. Figure 6.3 shows the output from the mixed model. Figure 6.4 shows the output from the fixed effects model.

The *anova()* function when applied to an lmer object (which we have called "mixed") provides F values for the fixed effects included in the model, but does not compute the *p*-values. The numerator degrees of freedom for a discrete effect is the number of levels minus one. The denominator degrees of freedom is the degrees of freedom for the residuals, which is the total sample size minus the total number of parameters in the model (including the intercept). Continuously valued effects always have one degree of freedom. Thus, in this model, there are five fixed effects, each having one degree of freedom, and two random effects (one of which, Dyad, is nested in the other). The degrees of freedom for these two random effects together is number of nests − 1 + number of nests *(number of Dyads per nest − 1) = 18 − 1 + 18*(1) = 35. Thus, the degrees of freedom for the residuals is 72 − 6 − 35 = 31. There are six fixed effect parameters, including the intercept. So, to find the *p*-value of the Treatment effect, use the R function *pf()*:

> pf(q = 103.6631, df1 = 1, df2 = 31, lower.tail=FALSE)
[1] 2.081934e-11

```
setwd("/Users/scottpardo/Desktop/SMABFS/programs & data/")
df1 <-read.csv("20161115 Example 6.2 Swallow Nestlings All
 Variables.csv")

attach(df1)
fNest <-factor(Nest)
fDyad <-factor(Dyad)

mixed <-lmer(WG ~ Treatment + Sex + Treatment:Sex + Theta + Upsilon +
 rA + (1|fNest/fDyad))
fixed.part <-fixef(mixed)
stderror <-diag(vcov(mixed))

anova(mixed)

summary(mixed)

fixed.model <-lm(WG ~ Treatment + Sex + Treatment:Sex + Theta +
 Upsilon + rA + fNest + fDyad%in%fNest)
```

FIGURE 6.2
R code for mixed model: Several fixed and random effects.

```
> anova(mixed)
Analysis of Variance Table
             Df       Sum Sq       Mean Sq       F value
Treatment    1        0.94990      0.94990       103.6631
Sex          1        0.77791      0.77791        84.8931
Theta        1        0.45800      0.45800        49.9812
Upsilon      1        1.87494      1.87494       204.6133
rA           1        0.01357      0.01357         1.4811
> summary(mixed)
Linear mixed model fit by REML ['lmerMod']
Formula: WG ~ Treatment + Sex + Theta + Upsilon + rA + (1 | fNest/fDyad)

REML criterion at convergence: -90.1

Scaled residuals:
   Min        1Q       Median       3Q        Max
-1.71118  -0.55704  -0.05133   0.58695   2.09197

Random effects:
 Groups        Name          Variance     Std.Dev.
 fDyad:fNest   (Intercept)   0.005262     0.07254
 fNest         (Intercept)   0.000000     0.00000
 Residual                    0.009163     0.09573
Number of obs: 72, groups:  fDyad:fNest, 36; fNest, 18


Fixed effects:
                  Estimate     Std. Error     t value
(Intercept)        0.43565      0.05931         7.345
TreatmentLight    -0.13699      0.03592        -3.813
SexMale            0.27993      0.03370         8.305
Theta             -0.98673      0.18163        -5.433
Upsilon           -1.99513      0.13898       -14.356
rA                 0.28194      0.23166         1.217

Correlation of Fixed Effects:
             (Intr)    TrtmnL    SexMal    Theta    Upsiln
TretmntLght  -0.326
SexMale      -0.120    0.027
Theta        -0.261   -0.583    -0.131
Upsilon       0.540    0.052     0.118    -0.102
rA           -0.681    0.609    -0.081    -0.179   -0.086
```

FIGURE 6.3
Output from mixed effects model.

```
> anova(fixed.model)
Analysis of Variance Table

Response: WG
                  Df   Sum Sq   Mean Sq   F value    Pr(>F)
Treatment          1   0.94990  0.94990   102.4811   2.390e-11 ***
Sex                1   1.67140  1.67140   180.3208   1.839e-14 ***
Theta              1   0.53438  0.53438    57.6526   1.466e-08 ***
Upsilon            1   2.59949  2.59949   280.4487   < 2.2e-16 ***
rA                 1   0.03992  0.03992     4.3065   0.046353 *
fNest             16   0.10605  0.00663     0.7151   0.758154
Treatment:Sex      1   0.00309  0.00309     0.3329   0.568120
fNest:fDyad       18   0.55102  0.03061     3.3027   0.001709 **
Residuals         31   0.28734  0.00927
---
Signif. codes:  0 '***' 0.001 '**' 0.01 '*' 0.05 '.' 0.1 ' ' 1
```

FIGURE 6.4
Output from fixed effects model.

Keep in mind that F-statistics are large when the null hypothesis of no effect is false. Thus, the upper tail probability for the F value is the *p*-value.

The same logic applies to F values from the purely fixed effects model. The difference lies in the way in which the residual sums of squares is computed in the mixed model. As in the case of the *lme()* function, the default methodology used for *lmer()* is REML, whereas in fixed effects models, the methodology is least squares. In the case of these data, the differences were fairly small.

Although *p*-values can be instructive as to the relevance of regressors (compute the *p*-value for the rA term in the mixed model), perhaps more importantly is the ability of the model to "predict" the response. To obtain a list of *BLUP* predictions from the *lmer* function, use the function *fitted()* (sometimes referred to as a "method" in R parlance) together with the object "mixed":

> fitted(mixed)

In fact, we can plot the *BLUP* and *BLUE* values against the actual observations, as shown in Figure 6.5. In this case, it does not appear that one model produces better predictions than the other. This may not always be the case.

Figure 6.6 is an illustration of barn swallows.

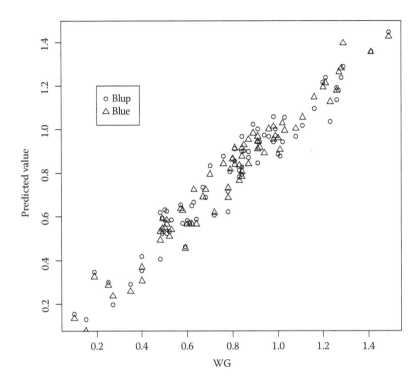

FIGURE 6.5
BLUE and *BLUP* plotted against actual observed weight gain.

FIGURE 6.6
Barn swallows (*Hirundo rustica*).

Theoretical Aspects

One use of mixed/random effects models is to make a guess at the value of the response for each individual in the dataset. Recall that the model is

$$WG_{ijk} = \mu + \tau_i + \gamma_j + \varepsilon_{ijk}$$

We are assuming that the observed value of weight gain in an individual is a departure from the overall average weight gain, where the departure is deterministically altered by the treatment (Dark or Light underside), the particular dyad to which the nestlings belong, and some random noise. While WG_{ijk} is the actual observed value, we recognize that since dyad is a random effect, and of course there is noise, perhaps the guess about the individual's weight gain should be somehow adjusted. As mentioned earlier, such an adjustment is called the best linear unbiased predictor, or *BLUP* (Searle, op.cit.). In this case, the *BLUP* for the effect of being in a particular dyad on weight gain is

$$\hat{\gamma}_{ijk} = \frac{n_j \hat{\sigma}_D^2}{\hat{\sigma}_\varepsilon^2 + n_j \hat{\sigma}_D^2} \left(WG_{ijk} - \widehat{\left(\mu + \tau_i \right)} \right)$$

The value of n is the number of values observed for each dyad, which in this case was $n_j = n = 2$ for all dyads. The term $\mu + \tau_i$ represents the average value of *WG* for Treatment *i*. Suppose $i = 1$ for Light, and $i = 2$ for Dark. For Light underside, the value of *WG* is Intercept + TreatmentLight ≈ 0.9075 − 0.2297 = 0.6778. The observed weight gain in dyad #1 under the "Light" Treatment was 0.78. The multiplier, also called the shrinkage (because it is a value less than one) is

$$\frac{n \hat{\sigma}_D^2}{\hat{\sigma}_\varepsilon^2 + n \hat{\sigma}_D^2} \approx 0.4993$$

So the *BLUP* for the effect of being in dyad #1 with "Light" Treatment is

$$BLUP_{1,1} = 0.4993 \left(0.78 - 0.6778 \right) \approx 0.05103$$

Thus, the *BLUP* for weight gain in the individual in dyad #1 with the Light Treatment is

$$0.6778 + 0.05103 = 0.72883.$$

Clearly this *BLUP* is not the same as the observed value of *WG*, which was 0.78 for the individual in question. Thus, the *BLUP* was "shrunk" closer to the overall average WG for individuals with the Light underside Treatment. The *lme*() function can produce an object (e.g., mixedf and mixedr). One of the attributed of an *lme*() object is called *fitted*. These fitted values are not *BLUP*s. Rather, they are *BLUE*s, or best linear unbiased estimators. They are related to *BLUP*s, but they are not "shrinkage" estimators. The *BLUE* is the combination of coefficients estimated for the model. The *lme*() object has an attribute called *coefficients*. For example, the *BLUE* for dyad #1, Light underside Treatment individual is

$$\text{Intercept} + \text{TreatmentLight} + \text{dyad} \#1, \text{ Light individual}$$
$$\approx 0.9075 - 0.2297 - 0.05624 = 0.62156$$

Key Points for Chapter 6

- A random effect is an effect on a response variable by a factor or regressor whose values are sampled from a random variable.
- Models can have both random effects and fixed effects, or regressors whose values are not sampled from random variables.
- Mixed models have parameters whose values must be estimated; the method of choice is called restricted maximum likelihood (REML).
- Random effects can be thought to contribute to variability of the response; the part of the response variance due to each random effect can be estimated via REML.
- Calculating fitted or predicted values of the response in the face of random effects is achieved by using best linear unbiased predictors (*BLUP*).
- Calculating fitted or predicted values of the response in the face of purely fixed effects is achieved by using best linear unbiased estimators (*BLUE*).
- It is sometimes not necessarily clear when effects should be treated as random or fixed; the objective may be to obtain the best predictors; the decision to treat effects as fixed or random might be aided by observing the difference in magnitude of the average residuals between two different models, one treating the effect as fixed and the other treating the effect as random.

Exercises and Questions

1. Explain the difference between a fixed effect and a random effect; are there circumstances where it would make more sense to identify an effect as one or the other?

7

Polytomous Discrete Variables:
$R \times C$ Contingency Tables

General Ideas

Independence of Two Discrete Variables

In Chapter 2, we introduced the 2×2 table for investigating the relationship between two discrete variables with only two states each. If one or more variable polytomous, the analysis can proceed in a similar fashion. Instead of having a table with two rows and two columns, there would be R rows and C columns, where R is the number of states for the row variable, and C is the number of states for the column variable. These analyses are sometimes referred to as $R \times C$ contingency tables (Conover, 1980). They are also referred to as a chi-squared test, since a commonly used test statistic is assumed to have a chi-squared distribution under the null hypothesis of independence.

Examples with R Code

Male side-blotched lizards (*Uta stansburiana*) exhibit three different throat color morphs, each associated with a different sexual strategy. Orange-throated males are highly aggressive and defend a large territory with multiple females. Blue-throated males defend a smaller territory with fewer females, but can monopolize their females more effectively than the orange males because they have less area to defend. Yellow-throated males do not defend a territory at all, and instead attempt to sneak copulations on the territories of other males. Orange males outcompete blue males, blue males outcompete yellow males, and yellow males outcompete orange males, such that the relative frequencies of each morph within a population tend to cycle over time (Sinervo and Lively 1996). Bleay et al. (2007) investigated how the reproductive success of each morph is affected by its relative frequency in the population and by population density. They removed all the lizards from six rocky outcrops, and then reintroduced lizards to each outcrop to create six different populations with different color morph ratios and population densities. As an example, consider the following modified version of Bleay et al.'s original study. For the sake of simplicity, we will ignore the effects of population density, and will have four populations instead of six. The four populations differ in the relative frequencies of male color morphs, as follows:

Population 1: High frequency of orange males

Population 2: High frequency of blue males

Population 3: High frequency of yellow males

Population 4: Equal frequencies of all three morphs

The researchers genotype all the males in each population and all the offspring produced over the course of one breeding season in order to determine the total number of offspring sired by males of each morph in each population. Table 7.1 shows the frequency distributions of offspring by color morph and population.

The question is whether the distributions of offspring by male color morph differ between populations.

The null hypothesis is that the distributions of offspring by male color morphs are independent of the particular population. First, if color morph success is completely independent of population, then

$$Pr\left\{Offspring \mid Morph_i, \ Pop_j\right\} = Pr\left\{Offspring \mid Morph_i\right\}Pr\left\{Offspring \mid Pop_j\right\}$$

Thus, the expected number of offspring under the null hypothesis of independence would be

$$E[n_{ij}] = Pr\left\{Offspring \mid Morph_i\right\}Pr\left\{Offspring \mid Pop_i\right\} * N$$

The value of $Pr\{Offspring \mid Morph_i\}$ is estimated as the proportion of the $N = 330$ results that are in row i, and $Pr\{Offspring \mid Pop_j\}$ is estimated as the proportion that are in column j.

Morph is either Orange, Blue, or Yellow, and *Pop* is either Pop1, Pop2, Pop3, or Pop4. The n_{ij} represents the number of offspring by Morph i in Population j.

Table 7.2 shows the table of expected numbers in each "cell" of the table.

TABLE 7.1

Side-Blotched Lizard Offspring

	Population				Row	
Morph	Population 1	Population 2	Population 3	Population 4	Total	Row %
Orange males	51	41	23	28	143	43.33%
Blue males	15	20	36	27	98	29.70%
Yellow males	16	20	25	28	89	26.97%
Column Total	82	81	84	83	330	100.00%
Column %	24.85%	24.55%	25.45%	25.15%		

TABLE 7.2

Expected Numbers of Offspring

			Population			
Morph	Population 1	Population 2	Population 3	Population 4	Row Total	Row %
Orange males	36	35	36	36	143	43.33%
Blue males	24	24	25	25	98	29.70%
Yellow males	22	22	23	22	89	26.97%
Column Total	82	81	84	83	330	100.00%
Column %	24.85%	24.55%	25.45%	25.15%		

We rendered the numbers in Table 7.1 into a data file, and used the *chisq. test()* function to perform the analyses. The parameter simulate.p.value=TRUE generates the *p*-value using a permutation test, with parameter B indicating the number of permutations to attempt. Figure 7.1 shows the R code together with the associated output in the R console window.

In this example, the permutation test was used, with B = 2000 replicate random permutations selected. The total number of permutations possible is (3*4)! = 12! = 479,001,600. When the permutation test option is selected (simulate.p.value=TRUE) then the degrees of freedom parameter is no longer applicable, since the null distribution for the test statistic "X-squared" is derived empirically. If the χ^2 distribution were used to compute the *p*-value, then $df = (4 - 1)*(3 - 1) = 6$ and the output would be

> chisq.test(lizmat, simulate.p.value = FALSE)

Pearson's chi-squared test:

data: lizmat
X-squared = 27.317, df = 6, *p*-value = 0.0001263
Direct computation of the *p*-value using *pchisq*(q=27.317,df=6,lower. tail=FALSE) yields the value 0.0001262858.

The permutation test is conservative inasmuch as the permutation *p*-value is greater than the χ^2 distribution *p*-value. Nevertheless, both tests lead to the same conclusion; namely, the null hypothesis that offspring distributions by Morph and Population are independent of each other is rejected.

If the data are not arranged in a table, then an *R × C* table can be constructed using the *table()* function. If, for example, the data consisted of two columns, one indicating Morph and the other indicating Population, and each row of

```
setwd("C:\\Users\\SMFSBE\\Statistical Data & Programs")
df1 <-read.csv("20170926 Example 7.1 Lizards.csv")
#
#
# VARIABLES:
# Morph
#   Pop1
#   Pop2
#   Pop3
#   Pop4

#
attach(df1)

lizmat <-data.frame(Pop1,Pop2,Pop3,Pop4,row.names=Morph)
chisq.test(lizmat,simulate.p.value=TRUE,B=2000)


OUTPUT

Pearson's Chi-squared test with simulated p-value (based on 2000
    replicates)

data:  lizmat
X-squared = 27.317, df = NA, p-value = 0.0004998

> lizmat
        Pop1 Pop2 Pop3 Pop4
Orange    51   41   23   28
Blue      15   20   36   27
Yellow    16   20   25   28
```

FIGURE 7.1
R code for chi-squared test of independence.

the dataset indicated a particular offspring of a particular Morph in a particular Population, the code shown in Figure 7.2 would yield the same results as obtained with the code in Figure 7.1.

As another example, consider the following problem in animal coloration. The need for camouflage from predators is an important selection pressure in the evolution of animal color patterns, and different color patterns often evolve, in part, to help animals hide in different environments. In species with color polymorphism, microhabitat preferences may explain the existence of more than one color pattern within the same population. The pygmy grasshopper (*Tetrix undulata*) has multiple color morphs, including solid black, solid light gray, and dark-and-light striped (among other colors). Ahnesjö and Forsman (2006) tested whether pygmy grasshoppers prefer microhabitats that match their own color pattern by placing grasshoppers in an arena with multiple different microhabitats and recording their position every five minutes for one hour. Here, we consider a modified version of their experiment.

Thirty-three black grasshoppers, twenty-four gray grasshoppers, and twenty-five striped grasshoppers were tested in an arena containing three different substrates: Light-colored spruce needles, dark-colored spruce needles, and a mix of light and dark spruce needles. The grasshoppers were placed in the center of the arena, and the amount of time they spent in each substrate sector was recorded. The sector in which an individual

```
setwd("C:\\Users\\SMFSBE\\Statistical Data & Programs")
df1 <-read.csv("20170926 Example 7.1 Lizards Alt.csv")
#
#
# VARIABLES:
#   Morph
#   Pop
#
attach(df1)
lizmat <-table(df1[,c("Morph","Pop")])

chisq.test(lizmat,simulate.p.value=TRUE,B=2000)

OUTPUT
Pearson's Chi-squared test with simulated p-value (based on 2000
     replicates)

data:  lizmat
X-squared = 57.58, df = NA, p-value = 0.0004998

> lizmat
          Pop1 Pop2 Pop3 Pop4
Orange    51   41   23   28
Blue      15   20   36   27
Yellow    16   20   25   28
```

FIGURE 7.2
R code for chi-squared test of independence: Alternative data format.

grasshopper spent the most time was scored as its preferred substrate color. The researchers hypothesized that the grasshoppers would prefer the substrate color that most closely matches their own color pattern in order to facilitate camouflage. Under this hypothesis, they predicted that the black grasshoppers would prefer the dark needles, the gray grasshoppers would prefer the light needles, and the striped grasshoppers would prefer the mixed needles. The null hypothesis is that substrate color is independent of color morph. If this were true, we would expect to see an equal proportion of individuals from each morph preferring each substrate color.

The analysis and the code for Example 7.2 is almost identical to that of Example 7.1, except that in this case, the null hypothesis is that the probability for a given color morph to choose any substrate color is uniform. Since there were three substrate colors, the expected fraction of times a given color morph would choose a given substrate is 1/3, or about 33.33 percent. The added parameter in the *chisq.test*() function, p, is a vector of null probabilities. The assignment

$$\text{ptest} <- \text{rep}(1/\text{length(morph)}, \text{length(morph)})$$

creates the vector of null probabilities. Figure 7.3 shows the code and output.

```
setwd("C:\\Users\\Failsafe\\Statistical Methods for Animal Behavior Field Studies
\\Statistical Data & Programs")
df1 <-read.csv("20161128 Example 7.2 Grasshopper Color Morphs.csv")
#
#
# VARIABLES:
#   Dark
#   Light
#   Mixed

#
attach(df1)

morph <-data.frame(Dark,Light,Mixed,row.names=c("Black","Gray","Striped"))

ptest <-rep(1/length(morph),length(morph))

chisq.test(morph,p=ptest,simulate.p.value=TRUE,B=2000)


OUTPUT

> morph <-data.frame(Dark,Light,Mixed,row.names=c("Black","Gray","Striped"))
> ptest <-rep(1/length(morph),length(morph))
> chisq.test(morph,p=ptest,simulate.p.value=TRUE,B=2000)

    Pearson's Chi-squared test with simulated p-value (based on 2000
    replicates)

data:  morph
X-squared = 16.185, df = NA, p-value = 0.002499

> morph
        Dark   Light   Mixed
Black     5      7      21
Gray     14      5       5
Striped   7      9       9
```

FIGURE 7.3
R code for grasshopper/substrate color table.

A Goodness-of-Fit Test

Sometimes a researcher would like to know if certain assumptions about the underlying distribution of some variable are justifiable. A Chi-squared statistic like X^2 can be used to decide whether continuously valued data seem to come from a particular type of distribution, say, normal. The idea is to first partition the data into discrete, nonoverlapping intervals, which we will call *bins*. Then count the number of observations in each bin, and compute the expected number of observations in each bin based on the hypothesized distribution.

As an example, consider the body weight of the shrimp, *Gammarus duebeni*. Suppose the weights for N = 60 shrimp were obtained, and the histogram looks like that shown in Figure 7.4.

The *hist*() function can be used to determine the frequency (counts) for each bin. To test for normality, the expected number of observations out of N = 60 can be computed using the *pnorm*() function. Figure 7.5 shows the R code that produced the histogram and subsequently computed the *p*-value for the χ^2 test statistic ("X-squared" in the console output).

Based on the permutation test *p*-value of 0.6577, the hypothesis of normality would not be rejected. Another goodness-of-fit test is called the

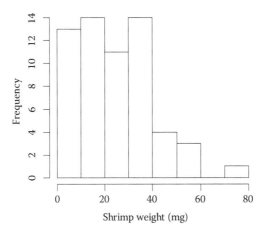

FIGURE 7.4
Histogram of shrimp weight.

```
setwd("C:\\Users\\Failsafe\\Statistical Methods for Animal Behavior Field Studies\\Statistical
Data & Programs")
df1 <-read.csv("20161129 Example 7.3 Shrimp Weight.csv")
#
#
# VARIABLES:
#   ShrimpWt
#
#
attach(df1)
expected <-c()
observed <-c()

nsize <-length(ShrimpWt)
muhat <-mean(ShrimpWt)
sdhat <-sd(ShrimpWt)
htest <-hist(ShrimpWt,xlab="Shrimp Weight(mg)")

nbins <-length(htest$breaks)

for (i in 2:nbins) {

 expected[i-1] <-nsize*(pnorm(q=htest$breaks[i],mean=muhat,sd=sdhat) -
pnorm(q=htest$breaks[i-1],mean=muhat,sd=sdhat))

 observed[i-1] <-htest$counts[i-1]
 }

RxC_table <-cbind(expected,observed)

chisq.test(RxC_table,simulate.p.value=TRUE,B=2000)

OUTPUT
> chisq.test(RxC_table,simulate.p.value=TRUE,B=2000)

    Pearson's Chi-squared test with simulated p-value (based on 2000
    replicates)

data:  RxC_table
X-squared = 4.8585, df = NA, p-value = 0.6577
```

FIGURE 7.5
Histogram and goodness-of-fit test code for shrimp weight data.

Kolmogorov–Smirnov test (Conover, 1980). The R function *ks.test()* implements the computations. There are two basic alternatives: One compares the empirical distributions of two variables, and the other will compare a single variable's empirical distribution to some parametric form (e.g., normal). Thus, to compare the shrimp weight distribution to a normal, one could use the code:

$$ks.test(\text{ShrimpWt,"pnorm",mean(ShrimpWt),sd(ShrimpWt)})$$

That is, the parametric form is normal (the pnorm function) with μ = mean(ShrimpWt) and σ = sd(ShrimpWt). The output is

> ks.test(x=ShrimpWt,"pnorm",mean(ShrimpWt),sd(ShrimpWt))

One-sample Kolmogorov–Smirnov test:

data: ShrimpWt
D = 0.094903, p-value = 0.6523
alternative hypothesis: two-sided

Warning message:
In ks.test(x = ShrimpWt, "pnorm", mean(ShrimpWt), sd(ShrimpWt)):
  ties should not be present for the Kolmogorov–Smirnov test

Thus, as in the case of the χ^2 statistic, the Kolmogorov–Smirnov statistic fails to reject the null hypothesis that the data, ShrimpWt, come from a normally distributed population.

A word of caution: The data were actually simulated from a gamma distribution, which is decidedly not normal. The histogram may not appear very symmetric about the mean weight. The moral is to trust your instincts, and not to surrender to a *p*-value. Also, remember that *p*-values from permutation and other nonparametric tests have fewer assumptions that may be violated.

A Special Goodness-of-Fit Test: Test for Random Allocation

E. O. Wilson (1980) identified four physical castes of leaf-cutter ant (*A. sexdens*) workers based on body size (measured as head width), and observed that

TABLE 7.3

Leaf-Cutter Ant Size Castes

Caste	Head Width (mm)
Caste 1	1.0
Caste 2	1.4
Caste 3	2.2
Caste 4	3.0

different castes specialize on different behavioral tasks. Imagine a study similar to Wilson's, in which a researcher identifies 19 different categories of behavior and counts up the number of workers in each size class that performed each behavior. The first question that he asks is whether physical caste is significantly associated with behavior. This question is simply answered by the independence test already discussed. The researcher then wants to know whether each individual caste is associated with certain behaviors more frequently than expected by random chance. This second question can be answered by testing whether the occurrence of different behaviors are uniformly distributed within a given caste. The size classes, or "castes," are listed in Table 7.3.

The head width classification for castes is of course a "middle" value for some range. For our purposes, we will treat those designations as absolute. Figure 7.6 shows the output from R code.

```
> df2 <-cbind(Caste1.1.0mm,Caste2.1.4mm,Caste3.2.2mm,Caste4.3.0mm)
> chisq.test(df2,simulate.p.value=TRUE,B=2000)

    Pearson's Chi-squared test with simulated p-value (based on 2000
    replicates)

data:  df2
X-squared = 1891.6, df = NA, p-value = 0.0004998

>
>
> ptest <-rep(1/no.rows,no.rows)
> chisq.test(Caste1.1.0mm,p=ptest,simulate.p.value=TRUE,B=2000)

    Chi-squared test for given probabilities with simulated p-value (based
    on 2000 replicates)

data:  Caste1.1.0mm
X-squared = 596.26, df = NA, p-value = 0.0004998

> chisq.test(Caste2.1.4mm,p=ptest,simulate.p.value=TRUE,B=2000)

    Chi-squared test for given probabilities with simulated p-value (based
    on 2000 replicates)

data:  Caste2.1.4mm
X-squared = 216.59, df = NA, p-value = 0.0004998

> chisq.test(Caste3.2.2mm,p=ptest,simulate.p.value=TRUE,B=2000)

    Chi-squared test for given probabilities with simulated p-value (based
    on 2000 replicates)

data:  Caste3.2.2mm
X-squared = 523.85, df = NA, p-value = 0.0004998

> chisq.test(Caste4.3.0mm,p=ptest,simulate.p.value=TRUE,B=2000)

    Chi-squared test for given probabilities with simulated p-value (based
    on 2000 replicates)

data:  Caste4.3.0mm
X-squared = 552.28, df = NA, p-value = 0.0004998
```

FIGURE 7.6
Leaf-cutter ants caste behavior distributions output.

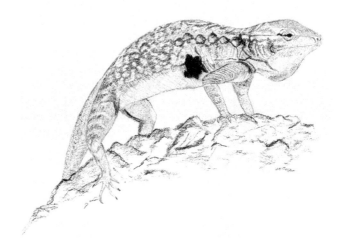

FIGURE 7.7
Side-blotched lizards (*Uta stansburiana*).

Other questions of interest might be, "What is the probability of observing a larval feeding (LAF) behavior in Caste 3 leaf-cutters?" That type of question will be answered in the chapter on Bayesian inference.

Figure 7.7 is an illustration of the side-blotched lizard.

Theoretical Aspects

Karl Pearson (1900) described a method for deciding whether two variables were independent or correlated. The method involved partitioning the range of values for each variable into discrete, nonoverlapping bins, and counting up the numbers of observations in each cell, or bin, for one variable crossed with the bins of the other. If O_{ij} represents the number of observations in row i, column j, and E_{ij} represents the expected number of results in row i and column j (under the hypothesis that row and column variables are independent), then the statistic

$$X^2 = \sum_{i=1}^{R}\sum_{j=1}^{C}\frac{(O_{ij}-E_{ij})^2}{E_{ij}}$$

has (approximately, at least) a chi-squared distribution with $(R-1)*(C-1)$ degrees of freedom (a parameter of the χ^2 distribution). R = number of rows (4), and C = number of columns (3). Thus, a p-value for the statistic can be found using the R function *pchisq()*. Fortunately, the function *chisq.test()* will do the analyses on raw data.

The Kolmogorov–Smirnov test statistic is the maximum absolute value of difference between the CDFs of the two distributions being compared, whether it is two sample distributions or a sample distribution compared to some hypothetical distribution. The statistic has a particular distribution, based on combinatorial methods, under the null hypothesis that the two distributions are the same. Thus, the *p*-value is the tail probabilities (usually two-sided) from the null distribution.

Key Points for Chapter 7

- Multiple discrete variables may be observed on a single experimental unit.
- The array of frequencies of two discrete variables in a table is called a *contingency table*.
- One question often asked is whether the values of one variable are statistically independent of the other variable.
- It is possible to calculate a statistic that under some assumptions has a χ^2 distribution under the hypothesis that the two variables are in fact independent.
- A test can be performed to yield a *p*-value either using the χ^2 assumption or via a permutation test.
- Make graphs to visualize data, and trust your instincts.

Exercises and Questions

1. Is it better to use a permutation test or rely on the conventional chi-squared test statistic and associated *p*-value for an R × C contingency table?

8

The Generalized Linear Model: Logistic and Poisson Regression

General Ideas

Binary Logistic Regression

If P represents the probability of success, then $1 - P$ is the probability of "not success." The odds of success are

$$O = \frac{P}{1-P}$$

As an example, suppose that we are interested in the odds that an individual in a population has trait "X." Furthermore, suppose that the probability of an individual having trait "X" is thought to be

$$P = 0.50 = \frac{50}{100}$$

and

$$1 - P = 0.50 = \frac{50}{100}$$

The odds that an individual has trait "X" is then:

$$O = \frac{P}{1-P} = \frac{50/100}{50/100} = \frac{50}{50} = 1$$

Thus, the odds of "X" versus not "X" are 1, which is called *even odds*. That is, there is no greater or lesser likelihood that an individual has trait "X" versus not "X." Suppose we consider a different trait, "Y," thought to have approximately a 66.66 percent prevalence. Then we have:

$$P = 0.6666$$

$$1 - P = 0.3334$$

$$O = \frac{P}{1-P} = \frac{0.6666}{0.3334} = 1.9994 \approx 2$$

Now we say that it is two times as likely that an individual has trait "Y," versus not, or the odds that an individual has "Y" are two to one in favor. Now you know that if you are told that the odds are three to one in favor of a particular horse to win a race, you know that it is three times as likely that this horse will win, versus not.

An odds ratio is the ratio of odds for two different conditions. Suppose there are two related species of *Plasmodium* (the parasitic protozoan that causes malaria), which differ in virulence (i.e., how likely they are to kill their host). Let P_1 represent the probability that species 1 will kill its host, and P_2 the probability that species 2 will kill its host. We will presume that the virulences of the two species are independent of one another. Thus P_1 and P_2 are not necessarily complementary; P_1 is not necessarily equal to $1 - P_2$. The odds of each species being fatal to its host are

$$O_1 = \frac{P_1}{1 - P_1}$$

$$O_2 = \frac{P_2}{1 - P_2}$$

The odds ratio is then

$$\frac{O_1}{O_2} = \frac{P_1/(1 - P_1)}{P_2/(1 - P_2)}$$

The odds ratio is the number of times more likely that the "numerator" species is to kill its host compared to the "denominator" species. Suppose $P_1 = 0.90$ and $P_2 = 0.45$. Then the odds ratio of species 1 to species 2 is

$$\frac{O_1}{O_2} = \frac{P_1/(1 - P_1)}{P_2/(1 - P_2)} = \frac{0.90/0.10}{0.45/0.65} \approx 13$$

So species 1 is 13 times more likely to kill its host than species 2. The ratio of P_1 to P_2 is only 0.90 / 0.45 = 2. This ratio (P_1/P_2) does not incorporate the probability that an individual from a given species will *not* kill its host. By contrast, the odds ratio does account for the relative probability of "success" and "failure" for each of the two entities being compared.

Examples with R Code

The Logit Transformation

It will be convenient to use the natural logarithm of odds rather than the odds directly. So we will define the logit (pronounced "low-jit") of P to be

$$\lambda = \ln(O) = \ln\left(\frac{P}{1-P}\right)$$

If you are given the logit's value, you can solve for P:

$$P = \frac{1}{1+e^{-\lambda}}$$

Keep in mind that P represents the probability of obtaining a success; however, success is defined. Furthermore, we will define a Bernoulli random variable, Y, to have the following binary values:

$$Y = \begin{cases} 1 \text{ if "success"} \\ 0 \text{ otherwise} \end{cases}$$

Thus,

$$P = \Pr\{Y = 1\}$$

Suppose that the logit was in fact a linear function of some regressor, X, so that

$$\lambda(X) = \ln(O) = \ln\left(\frac{P(X)}{1-P(X)}\right) = \beta_0 + \beta_1 X$$

By solving for P(X), we get

$$P(X) = \Pr\{Y = 1 | X\} = \frac{1}{1 + e^{-\lambda(X)}} = \frac{1}{1 + e^{-(\beta_0 + \beta_1 X)}}$$

Once data are collected, the parameters β_0 and β_1 may be estimated. The estimates may be obtained in many ways, but a common way is via maximum likelihood.

The values of the x_j could be chosen in a designed experimental fashion, say, a 2^{k-p} fractional factorial experimental design (see Montgomery, 2001). The y_i are binary observations. For the moment, we will assume that the x_j are continuously valued regressors, and our objective is to find a predictive equation for the probability of a success ($Y = 1$) in terms of the x_j. Furthermore, in building this probability model, we would like to determine which if any of the regressors (x_j) actually affect the probability that $Y = 1$.

As an example, suppose that a researcher is studying foraging behavior in the straw-colored fruit bat (*Eidolon helvum*), a species known to travel long distances in search of food. She would like to predict the probability that an individual bat will travel to a particular fruiting tree as a function of the distance between the tree and the bat's roost, so she places GPS tags on some bats and records where they go. The R function *glm*() can be used to fit a binary logistic regression model. The code in Figure 8.1 shows an implementation of *glm*() together with significance tests and confidence intervals for the model parameters.

The *glm*() function produces a table of parameter estimates (in this case, an "intercept" and a coefficient for the single regressor, Distance), together with approximate standard errors for the estimates and p-values based on the assumption that the estimates are normally distributed, and the null hypothesis that the parameter is equal to zero. The *anova*() function allows the user to specify a chi-squared approximation instead of a normal approximation for computing p-values. Although the p-values for the normal and chi-squared approximations differ, they are both less than 0.05, so they would lead the researcher to the same conclusion. The conclusion, of course, is that the probability of a visit to the location is in fact related to Distance.

The statement

```
df1$probmod<-predict(model_bin,type="response")
```

produces a new column in dataframe df1 that contains the predicted probability of a group being present at the location of interest. Figure 8.2 shows a plot of the predicted probabilities by Distance.

Unlike the case of a continuous response, the response variable for binary logistic regression is, well, binary, whereas the predicted value is a decimal number between zero and one. Thus, for binary logistic regression it is not possible to plot predicted values against actual responses.

As an example of logistic regression with discrete "regressor" variables (factors), consider the work done by Saporito et al. (2017) on poison dart frogs. Poison dart frogs of the family Dendrobatidae exhibit brightly colored skin, which is thought to serve as an honest signal to predators of their toxicity (aposematism). Saporito and his colleagues sought to test whether predators do in fact avoid brightly colored frogs. They constructed 800 plastic models of frogs, and painted half to look like the brightly colored strawberry poison dart frog (*Oophaga pumilio*), while the remainder were painted brown. They then placed the models along transects in the forest. In order to determine whether camouflage also plays a role in determining predation rate, half of

```
setwd("C:\\Users\\SMFSBE\\Statistical Data & Programs")
df1 <-read.csv("20161130 Example 8.1 Logistic Regression One Regressor.csv")
# Variables
# Distance (km)
# Present (1 = Yes, 0 = No)
#
attach(df1)

model_bin <-glm(Present ~ Distance,family = binomial("logit"),na.action=na.omit)
#
# the function "attributes" will tell you the attributes of the glm object model_bin
#
# type: attributes(model_bin) at the > prompt in the R command window
#
# In particular, the attribute model_bin$fitted.values are the predicted probabilities that
# each observation should be classified into the group where Y = 1 (instead of Y = 0)
#
summary(model_bin)
confint(model_bin)
anova(model_bin,test="Chisq") #computes a sequential likelihood ratio chi squared statistic
#                    for each effect in the model
#
df1$probmod <-predict(model_bin,type="response")
#
# for glm objects the predict function defaults to returning log odds ratios
# the parameter type ="response" will make predict return predicted probabilities of a "1"
response
#


write.csv(df1,file="20161130 Example 8.1 Logistic Regression One Regressor OUTPUT.csv")

OUTPUT

Call:
glm(formula = Present ~ Distance, family = binomial("logit"),
   na.action = na.omit)

Deviance Residuals:
   Min      1Q    Median      3Q      Max
-1.4542  -1.0011  -0.6776  1.1506  1.8648

Coefficients:
         Estimate    Std. Error    z value    Pr(>|z|)
```

FIGURE 8.1
Logistic regression with one continuously valued regressor. *(Continued)*

```
(Intercept)              1.20053          0.81449          1.474    0.1405
Distance                -0.05757          0.02704         -2.129    0.0333 *
---
Signif. codes:  0 '***' 0.001 '**' 0.01 '*' 0.05 '.' 0.1 ' ' 1

(Dispersion parameter for binomial family taken to be 1)

    Null deviance: 66.406  on 49  degrees of freedom
Residual deviance: 60.672  on 48  degrees of freedom
AIC: 64.672

Number of Fisher Scoring iterations: 4

>confint(model_bin)
Waiting for profiling to be done...
                2.5 %              97.5 %
(Intercept) -0.3141611        2.933006092
Distance -0.1172384          -0.009624307
> anova(model_bin,test="Chisq") #computes a sequential likelihood ratio chi squared statistic
Analysis of Deviance Table

Model: binomial, link: logit

Response: Present

Terms added sequentially (first to last)


           Df              Deviance      Resid. Df    Resid. Dev    Pr(>Chi)
NULL       49              66.406
Distance   1               5.7346             48        60.672      0.01663 *
---
Signif. codes:  0 '***' 0.001 '**' 0.01 '*' 0.05 '.' 0.1 ' ' 1
> #                       for each effect in the model
```

FIGURE 8.1 (CONTINUED)
Logistic regression with one continuously valued regressor.

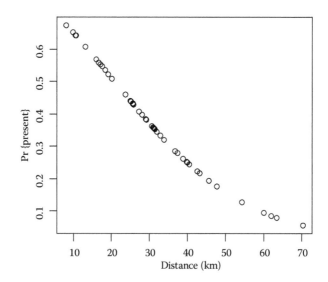

FIGURE 8.2
Predicted probabilities by distance.

the models of each color were placed directly on the brown leaf litter, while half were placed on a piece of white paper. The researchers collected the models after 48 hours, and assessed whether they had been predated upon based on whether they bore marks from beaks or teeth. In our example, based on the work of Saporito et al., there are 60 "model" frogs of each color on each type of background (total of 4*60 = 240 models). The regressors are model color (red or brown) and background (white or brown leaf-litter). In our simulated example, the logit model is symbolically:

$$\lambda(X) = Frog + Background + Frog * Background$$

The vector variable X represents coded values that are used to estimate the effects of the factors Frog and Background, and their interaction.

Figure 8.3 shows the R code and associated console output for fitting the logistic model to the data.

```
setwd("C:\\Users\\SMFSBE\\Statistical Data & Programs")
df1 <-read.csv("20170821 Example 8.2 Poison Dart Frogs.csv")
# Variables
# Index
# Frog (Red or Brown)
# Background (White or Leaf)
# Predate (1 = Yes 0 = No)
#
attach(df1)

model_bin <-glm(Predate ~ Frog + Background + Frog:Background,family =
binomial("logit"),na.action=na.omit)
#
# the function "attributes" will tell you the attributes of the glm object model_bin
#
# type: attributes(model_bin) at the > prompt in the R command window
#
# In particular, the attribute model_bin$fitted.values are the predicted probabilities that
# each observation should be classified into the group where Y = 1 (instead of Y = 0)
#
summary(model_bin)
confint(model_bin)
anova(model_bin,test="Chisq") #computes a sequential likelihood ratio chi squared statistic
#                               for each effect in the model
#
df1$probmod <-predict(model_bin,type="response")
#
# for glm objects the predict function defaults to returning log odds ratios
# the parameter type ="response" will make predict return predicted probabilities of a "1"
response
#

interaction.plot(x.factor=Background,trace.factor=Frog,response=df1$probmod,type="b",main=
"Predicted Predation Probability",xlab="Background",ylab="Pr{Predation}")
write.csv(df1,file="20170821 Example 8.2 Poison Dart Frogs OUTPUT.csv")

OUTPUT

Call:
glm(formula = Predate ~ Frog + Background + Frog:Background,
  family = binomial("logit"), na.action = na.omit)

Deviance Residuals:
  Min      1Q    Median      3Q      Max
```

FIGURE 8.3

Logistic regression with discrete factors: Poison dart frogs. *(Continued)*

```
-1.2059  -1.1213  -0.7585   1.2346   1.6651


Coefficients:

                         Estimate   Std. Error    z value   Pr(>|z|)
(Intercept)               -0.2683      0.2605      -1.030     0.3031
FrogRed                    0.3350      0.3669       0.913     0.3613
BackgroundWhite            0.1347      0.3672       0.367     0.7137
FrogRed:BackgroundWhite   -1.3000      0.5390      -2.412     0.0159 *
---
Signif. codes:  0 '***' 0.001 '**' 0.01 '*' 0.05 '.' 0.1 ' ' 1


(Dispersion parameter for binomial family taken to be 1)


    Null deviance: 326.01  on 239  degrees of freedom
Residual deviance: 315.61  on 236  degrees of freedom
AIC: 323.61


Number of Fisher Scoring iterations: 4
> confint(model_bin)
Waiting for profiling to be done...
                            2.5 %      97.5 %
(Intercept)             -0.7875304   0.2393226
FrogRed                 -0.3824734   1.0596757
BackgroundWhite         -0.5854786   0.8579550
FrogRed:BackgroundWhite -2.3681761  -0.2515239
> anova(model_bin,test="Chisq") #computes a sequential likelihood ratio chi squared statistic
Analysis of Deviance Table

Model: binomial, link: logit

Response: Predate

Terms added sequentially (first to last)


                Df    Deviance   Resid. Df   Resid. Dev   Pr(>Chi)
NULL                               239         326.01
Frog             1    1.0981      238         324.91      0.29469
Background       1    3.3844      237         321.53      0.06581
Frog:Background  1    5.9204      236         315.61      0.01497 *
---
Signif. codes:  0 '***' 0.001 '**' 0.01 '*' 0.05 '.' 0.1 ' ' 1
```

FIGURE 8.3 (CONTINUED)
Logistic regression with discrete factors: Poison dart frogs.

From both the default outoput from *glm*() and *anova*() functions, we see that the only significant effect at the 0.05 level was the interaction between model frog color and background type. Figure 8.4 shows an interaction plot for the estimated probability of predation.

The plot illustrates the reason for the significant interaction. While background color appears to have little effect on predation probability for brown frog models, it has a large effect on red frog models. This example is not only a good illustration of how logistic regression can be used as a sort of "anova" for assessing effects of discrete factors, it also illustrates the importance of interactions.

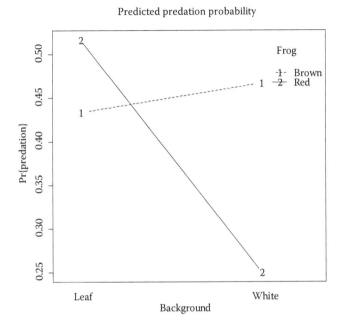

FIGURE 8.4
Interaction between frog and background color on predation probability.

Poisson Regression

Sometimes the response variable is a discrete count, such as the number of eggs in a clutch. One convenient model for count data is the Poisson distribution. If variable Y has a Poisson distribution, then the probability mass function is

$$Pr\{Y = y | \lambda\} = \frac{\lambda^y e^{-\lambda}}{y!}$$

The expected value of Y, $E[Y]$ is equal to λ.

The idea behind Poisson regression is that the logarithm of the parameter λ is a linear function of regressors $x = [x_0, x_1, x_2, \ldots, x_k]^T$:

$$\ln(E[Y|x]) = \ln(\lambda) = \boldsymbol{\beta}^T x$$

Where β is a vector of $k + 1$ unknown parameters, ($x_0 = 1$ yields the "intercept," β_0).

The R function *glm*() makes the whole process very easy to perform. When the response data are counts, and not zeroes and ones only, instead of using the family=binomial("logit"), use family=poisson("log"). As a simple example, suppose that instead of simply recording whether a fruit bat appeared at a particular tree, we recorded the number of visits to the tree as a function of the distance from the bat's roost. The code and output is shown in Figure 8.5.

Figure 8.6 shows a plot of the observed and predicted counts by distance.

As another example, consider the following study by Albo et al. (2013) on cryptic female choice in the nursery web spider (*Pisaura mirabilis*). In general, when females mate with more than one male, there are multiple routes through which they can exercise control over the paternity of their offspring. In addition to choosing with whom to mate in the first place, females who mate with multiple males may also preferentially use the sperm of certain mates over others when fertilizing their eggs internally

```
setwd("C:\\Users\\SMFSBE\\Statistical Data & Programs")
df1 <-read.csv("20161130 Example 8.3 Number of Visits.csv")
# Variables
# Distance (km)
# Count = No. of Visits
#
attach(df1)

model_pois <-glm(Count ~ Distance,family = poisson("log"),na.action=na.omit)
#
# the function "attributes" will tell you the attributes of the glm object model_pois
#
# type: attributes(model_pois) at the > prompt in the R command window
#
#
summary(model_pois)
confint(model_pois)
anova(model_pois,test="Chisq") #computes a sequential likelihood ratio chi squared statistic
#                 for each effect in the model
#
df1$predcount <-predict(model_pois,type="response")
#
# for glm objects with family=poisson("log"),
# the parameter type ="response" will make predict return predicted counts
#

plot(Distance,df1$predcount,type="n",main="Number of Visits by Distance",xlab="Distance
(km)",ylab="Number of Visits")
points(Distance,Count,pch=1,col=1)
points(Distance,df1$predcount,pch=2,col=4)
legend(x=50,y=16,legend=c("observed","predicted"),pch=c(1,2),col=c(1,4))
write.csv(df1,file="20161130 Example 8.3 Poisson Regression Number of Visits OUTPUT.csv")

OUTPUT
> summary(model_pois)

Call:
glm(formula = Count ~ Distance, family = poisson("log"), na.action = na.omit)

Deviance Residuals:
   Min       1Q    Median       3Q      Max
-1.5924  -0.7190  -0.1039   0.4344   2.1006
```

FIGURE 8.5

Poisson regression for number of visits.   *(Continued)*

```
Coefficients:
                 Estimate        Std. Error       z value         Pr(>|z|)
(Intercept)      3.060350          0.098458         31.08         <2e-16 ***
Distance        -0.020263          0.003273         -6.19         6e-10 ***
---
Signif. codes:  0 '***' 0.001 '**' 0.01 '*' 0.05 '.' 0.1 ' ' 1

(Dispersion parameter for poisson family taken to be 1)

    Null deviance: 78.483  on 49  degrees of freedom
Residual deviance: 37.021  on 48  degrees of freedom
AIC: 253.74

Number of Fisher Scoring iterations: 4

> confint(model_pois)
Waiting for profiling to be done...
                2.5 %          97.5 %
(Intercept)   2.86624081     3.25220903
Distance     -0.02675578    -0.01392406
> anova(model_pois,test="Chisq") #computes a sequential likelihood ratio chi squared statistic
Analysis of Deviance Table
Model: poisson, link: log
Response: Count
Terms added sequentially (first to last)
              Df      Deviance     Resid. Df      Resid. Dev      Pr(>Chi)
NULL          49       78.483
Distance       1       41.462          48           37.021       1.202e-10 ***
---
Signif. codes:  0 '***' 0.001 '**' 0.01 '*' 0.05 '.'      0.1 ' ' 1
> #                      for each effect in the model
> #
```

FIGURE 8.5 (CONTINUED)
Poisson regression for number of visits.

(cryptic female choice). *P. mirabilis* females mate with multiple males, and males often provide females with a nuptial gift (prey item wrapped in silk), which the female consumes during copulation. Females store sperm from males in specialized organs called spermathecae, and then use the sperm later to fertilize their eggs. As gift-giving may be an honest signal of male quality, Albo et al. (2013) conducted an experiment to test the hypothesis that females would store more sperm from males who provided a nuptial gift during copulation. In our example based on Albo et al.'s experiment, 128 virgin female spiders were randomly allocated to one of three groups: NG (N = 53), GT (N = 39), and G (N = 36). Females in the NG (no gift) group were mated to a male who did not have any nuptial gift to provide, while females in the other two groups were mated to a male with a nuptial gift. The difference between the GT and G groups was that in the GT (gift + terminated) group, the researchers terminated the copulation early to match the shorter duration of copulations observed in the NG group. In the G (gift) group, the researchers allowed the spiders to copulate undisturbed. After mating, the researchers dissected the

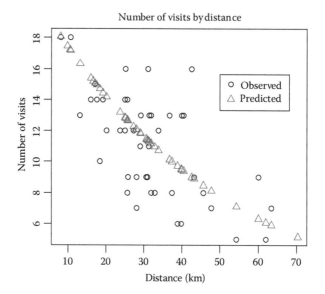

FIGURE 8.6
Observed and predicted counts by distance.

females and counted the number of sperm in their spermathecae under a microscope. In this example, the regressor is a discrete categorical variable, Gifting, with three levels. The response variable is Sperm, the number of sperm in the female's spermathecae. Figure 8.7 shows R code and console output for both Poisson regression and a normal-theory type one way ANOVA. The coefficients from the Poisson regression are the logarithms of the average value of Sperm for each "Gifting" treatment group. We use the *tapply()* function to display the means and variances of the data for each group. Inasmuch as the data are really Poisson distributed, the means and variances are "close" to each other for each group.

Figure 8.8 shows a box plot of the Sperm data by Gifting group.

Overdispersion

The Poisson distribution has a single parameter, λ, that is both its expected value and its variance. That is, for a Poisson random variable, K, with parameter λ:

$$E[K] = V[K] = \lambda$$

Sometimes count data seems like it ought to be Poisson, but their variance is greater than their mean. In such cases, we say the data are "overdispersed" (Gelman et al., 2000). A remedy for this situation is to use a discrete count

```
setwd("C:\\Users\\SMFSBE\\Statistical Data & Programs")
df1 <-read.csv("20170731 Example 8.4 Spider Sperm Count.csv")
# Variables
# ID
# Gifting
# Sperm
#
attach(df1)

model_norm <-lm(Sperm ~ Gifting -1)
model_pois <-glm(Sperm ~ Gifting -1,family = poisson("log"),na.action=na.omit)
#
# the function "attributes" will tell you the attributes of the glm object model_bin
#
# type: attributes(model_pois) at the > prompt in the R command window
#
#
anova(model_norm)
summary(model_pois)
confint(model_pois)
confint(model_norm)
tapply(X=Sperm,INDEX=Gifting,FUN=mean)
tapply(X=Sperm,INDEX=Gifting,FUN=var)

boxplot(Sperm ~ Gifting,main="Nursery Web Spider Sperm Count",xlab="Gifting
Group",ylab="Sperm Count")

OUTPUT

> anova(model_norm)
Analysis of Variance Table

Response: Sperm
            Df    Sum Sq    Mean Sq    F value    Pr(>F)
Gifting      3   1.3776e+10  4592064544  513403   < 2.2e-16 ***
Residuals  125   1.1180e+06     8944
---
Signif. codes:  0 '***' 0.001 '**' 0.01 '*' 0.05 '.' 0.1 ' ' 1
> summary(model_pois)

Call:
glm(formula = Sperm ~ Gifting-1, family = poisson("log"), na.action = na.omit)

Deviance Residuals:
   Min        1Q       Median       3Q        Max
-2.50159   -0.78870   -0.01529   0.68793   2.16747
```

FIGURE 8.7
Spider sperm count poisson regression: discrete regressor.

distribution model that has an extra parameter; the negative binomial is a natural extension of the Poisson, and is a candidate distribution for overdispersion models. Another alternative distribution is called *quasipoisson*. In the quasipoisson distribution, while $E[K] = \lambda$, the variance is some multiple of λ, with a multiplying parameter θ:

$$V[K] = \theta\lambda$$

The R function *glm()* allows the analyst to use the quasipoisson via the family input parameter. In this case, the link function, or linearizing transformation, is the log, as in the case of the ordinary Poisson family.

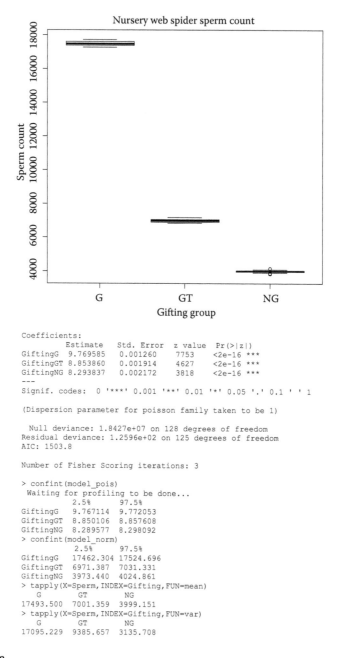

```
Coefficients:
           Estimate   Std. Error   z value   Pr(>|z|)
GiftingG  9.769585    0.001260     7753      <2e-16 ***
GiftingGT 8.853860    0.001914     4627      <2e-16 ***
GiftingNG 8.293837    0.002172     3818      <2e-16 ***
---
Signif. codes:  0 '***' 0.001 '**' 0.01 '*' 0.05 '.' 0.1 ' ' 1

(Dispersion parameter for poisson family taken to be 1)

   Null deviance: 1.8427e+07 on 128 degrees of freedom
Residual deviance: 1.2596e+02 on 125 degrees of freedom
AIC: 1503.8

Number of Fisher Scoring iterations: 3

> confint(model_pois)
 Waiting for profiling to be done...
           2.5%        97.5%
GiftingG   9.767114    9.772053
GiftingGT  8.850106    8.857608
GiftingNG  8.289577    8.298092
> confint(model_norm)
           2.5%        97.5%
GiftingG   17462.304   17524.696
GiftingGT  6971.387    7031.331
GiftingNG  3973.440    4024.861
> tapply(X=Sperm,INDEX=Gifting,FUN=mean)
    G          GT         NG
17493.500   7001.359    3999.151
> tapply(X=Sperm,INDEX=Gifting,FUN=var)
    G          GT         NG
17095.229   9385.657    3135.708
```

FIGURE 8.8
Box plot of sperm count data by gifting group.

Hinde (2006) conducted an experiment with great tits (*Parus major*), in which recordings of begging calls were played to parent birds, to observe how these affected the parents' feeding rates. She discovered that even when she only played the recordings to one of the two parents, the feeding rates of both parents increased, regardless of which parent was exposed to the playback. The following example is loosely based on Hinde's study. In this example, the response is the number of feeds per hour. There were three treatments: Playback to the female parent, playback to the male parent, and no playback. There were two discrete factors; namely, the treatment, and the sex of the feeding parent. Figure 8.9 shows R code, and associated output, for fitting a generalized linear model with a Poisson distribution family, and then with a quasipoisson. The two tables show the means and variances for all of the six groups of data.

The p-values from the chi-squared tests in the generalized ANOVA show that when ignoring the overdispersion, the feeding parent's sex was significant at the 0.05 level, as was the interaction between the treatment and the feeding parent's sex. However, when accounting for overdispersion, only the

```
setwd("C:\\Users\\SMFSBE\\Statistical Data & Programs")
df1 <-read.csv("20161205 Example 8.5 Overdispersed Poisson Feeds.csv")
# Variables
# Treatment: Control (no playback) PB_F (playback to female parent), PB_M (playback to male
parent)
# Feeder_Sex: F (female parent) M (male parent)
# Feeds: number of feedings in 1 hour
#
attach(df1)

#
#  First try a Poisson fit:
#
model_pois <-glm(Feeds ~ Treatment + Feeder_Sex + Treatment:Feeder_Sex,family =
poisson("log"),na.action=na.omit)

anova(model_pois,test="Chisq") #computes a sequential likelihood ratio chi squared statistic
#                               for each effect in the model
#
# Now try an "overdispersed" Poisson:
#
model_qpois <-glm(Feeds ~ Treatment + Feeder_Sex + Treatment:Feeder_Sex,family =
quasipoisson("log"),na.action=na.omit)
#
#
anova(model_qpois,test="Chisq") #computes a sequential likelihood ratio chi squared statistic
#                                for each effect in the model
# Check to see how close the mean and variance are
# for each experimental "condition"
#
tapply(Feeds,INDEX=list(Treatment,Feeder_Sex),FUN=mean)
tapply(Feeds,INDEX=list(Treatment,Feeder_Sex),FUN=var)

OUTPUT

>
> #
> #  First try a Poisson fit:
> #
> model_pois <-glm(Feeds ~ Treatment + Feeder_Sex + Treatment:Feeder_Sex,family =
poisson("log"),na.action=na.omit)
>
> anova(model_pois,test="Chisq") #computes a sequential likelihood ratio chi squared statistic
Analysis of Deviance Table
```

FIGURE 8.9
Ordinary and overdispersed Poisson models. *(Continued)*

```
Model: poisson, link: log

Response: Feeds

Terms added sequentially (first to last)

                      Df     Deviance      Resid. Df    Resid. Dev      Pr(>Chi)
NULL                  59     195.60
Treatment              2      72.648           57          122.96       < 2e-16 ***
Feeder_Sex             1       5.357           56          117.60       0.02064 *
Treatment:Feeder_Sex   2       7.506           54          110.09       0.02345 *
---
Signif. codes:  0 '***' 0.001 '**' 0.01 '*' 0.05 '.' 0.1 ' ' 1
> #          for each effect in the model
> #
> # Now try an "overdispersed" Poisson:
> #
> model_qpois <-glm(Feeds ~ Treatment + Feeder_Sex + Treatment:Feeder_Sex,family =
quasipoisson("log"),na.action=na.omit)
> #
> #
> anova(model_qpois,test="Chisq")#computes a sequential likelihood ratio chi squared statistic
Analysis of Deviance Table

Model: quasipoisson, link: log

Response: Feeds

Terms added sequentially (first to last)

                      Df     Deviance      Resid. Df   Resid. Dev   Pr(>Chi)
NULL                  59     195.60
Treatment              2      72.64857          57         122.96    8.83e-09 ***
Feeder_Sex             1       5.357            56         117.60    0.09818
Treatment:Feeder_Sex   2       7.506            54         110.09    0.14718
---
Signif. codes:  0 '***' 0.001 '**' 0.01 '*' 0.05 '.' 0.1 ' ' 1
> #          for each effect in the model
> # Check to see how close the mean and variance are
> # for each experimental "condition"
> #
> tapply(Feeds, INDEX=list(Treatment,Feeder_Sex),FUN=mean)
          F      M
Control 10.4   12.1
PB_F    23.6   17.4
PB_M    22.3   19.3

> tapply(Feeds, INDEX=list(Treatment,Feeder_Sex),FUN=var)
            F           M
Control 19.60000    21.87778
PB_F    16.93333    16.93333
PB_M    57.78889    72.90000
```

FIGURE 8.9 (CONTINUED)
Ordinary and overdispersed Poisson models.

treatment itself was significant. A clue to the need for the overdispersion model was that the variances are not equal to the average values for most of the experimental cases. If the overdispersion parameter of the quasipoisson $\theta = 1$, then we would expect the variances and means to be close in value to each other.

Zero-Inflated Data and Poisson Regression

Sometimes, count data can have a large preponderance of 0s, greater than the number expected for a Poisson distribution. A model can be fit to incorporate both the Poisson nature of the count data and the larger-than-expected numbers of zeros. Such models will be referred to as zero-inflated Poisson (ZIP) regressions. In ZIP models, the data appear to come in two groups: A group with a very high probability of zero counts, and another group that appears to have a Poisson distribution. The idea is that there are some individuals that would always yield a zero, regardless of the values of the regressors, and others where the probability of zero is based on the Poisson probability, which in turn depends on the values of the regressors.

Thus, a generalized regression, similar to the overdispersed Poisson, must be employed. Fortunately, R has a function called *zeroinfl()* in the package pscl, which will perform a zero-inflated Poisson regression.

Consider the following example based on the work of Schürch and Heg (2010) with the cichlid *Neolamprologus pulcher. N. pulcher* is a cooperatively breeding fish native to Lake Tanganyika. In this species, offspring may either disperse and attempt to breed on their own, or they may remain in their natal territory as nonbreeding "helpers" and assist their parents in raising their younger siblings. While Schürch and Heg studied the link between this life history decision and behavioral syndromes (personality) in *N. pulcher*, we will use a simplified example based on a single behavior to illustrate the methods for dealing with zero-inflated count data.

Suppose 60 young fish were assessed for their propensity to act as helpers. One of the behaviors that helpers routinely perform is to assist with maintenance of their group's territory. Thus, to measure propensity to help, the researchers shoveled sand into the group's shelter, and then counted the number of times each subject carried away a mouthful of the sand within a 10-minute period. As covariates/regressors, we will use the subject's body length and age. Figure 8.10 shows some R code for executing the *zeroinfl()* function, in package pscl, together with *summary()* output for OLS regression, Poisson regression, overdispersed Poisson regression, and zero-inflated Poisson regression. For the zero-inflated model, the algorithm requires starting values for model parameters, which may be provided by an "expectation-maximization" algorithm (the default). However, in some cases, data will require that the user supply starting values in order to have the zero-inflated model computations converge.

The only model for which all parameter estimates were significantly different from zero was the zero-inflated Poisson. Figure 8.11 shows a plot of predicted counts as a function of Length.

All models produce a sort of "average" prediction, and the predictions appear to fall somewhere in the middle of the actual observations. Only the zero-inflated model yielded significant p-values, at the $\alpha = 0.05$ level, for all the model parameters. Nevertheless, it is not clear which of these models produced the best predictions for the response.

```
setwd("C:\\Users\\SMFSBE\\Statistical Data & Programs")
df1<-read.csv("20170731 Example 8.6 Zero-Inflated Poisson Cichlid Helpfulness.csv")
#
#  need to execute library(pscl) once during R session
library(pscl)
#
# VARIABLES:
# Fish
# Length
# Age
# Carry
#
attach(df1)
linreg <- lm(Carry ~ Length + Age)
regpoisson <- glm(Carry ~ Length + Age,data=df1,family=poisson("log"))
overpoisson <- glm(Carry ~ Length + Age,data=df1,family=quasipoisson("log"))
#zeropoisson <- zeroinfl(Carry ~ Length + Age,control=zeroinfl.control(start=list(count=c(-
2,0.9,0.1),zero=c(-2, -0.2, -0.05))),data=df1,dist="poisson",link="log")
zeropoisson <- zeroinfl(Carry~Length + Age,EM=TRUE,data=df1,dist="poisson",link="log")
plot(Length,zeropoisson$fitted.values,pch=1,ylim=c(0,18),main="Poisson Models: Predicted vs.
Length")
points(Length,regpoisson$fitted.values,pch=2)
points(Length,overpoisson$fitted.values,pch=3)
points(Length, linreg$fitted.values,pch=4)
points(Length,Carry,pch=5)
legend(x=26,y=6,legend=c("Zero-Inflated Poisson","Regular Poisson","Overdispersed
Poisson","OLS Reg","Observed"),pch=c(1,2,3,4,5))
summary(linreg)
summary(regpoisson)
summary(overpoisson)
OUTPUT

Call:
lm(formula = Carry ~ Length + Age)

Residuals:
    Min      1Q    Median     3Q      Max
-15.2486  -2.8658  0.8821   4.0550   9.9959

Coefficients:
              Estimate   Std. Error   t value   Pr(>|t|)
(Intercept)   5.98021    16.42888     0.364     0.7172
Length        0.62225     0.32708     1.902     0.0622
Age          -0.06438     0.10332    -0.623     0.5357

---
Signif. codes:  0 '***' 0.001 '**' 0.01 '*' 0.05 '.' 0.1 ' ' 1
Residual standard error: 6.153 on 57 degrees of freedom
Multiple R-squared: 0.06712,    Adjusted R-squared: 0.03439
F-statistic: 2.051 on 2 and 57 DF,  p-value: 0.138

> summary(regpoisson)

Call:
glm(formula = Carry ~ Length + Age, family = poisson("log"),
  data = df1)

Deviance Residuals:
   Min      1Q    Median     3Q      Max
-5.5467  -0.8425  0.2225   1.0914   2.4431

Coefficients:
              Estimate   Std. Error   z value   Pr(>|z|)
(Intercept)   1.977750   0.732725     2.699     0.00695 **
Length        0.049512   0.015189     3.260     0.00112 **
Age          -0.004969   0.004573    -1.087     0.27724
```

FIGURE 8.10

Zero-inflated Poisson R script with output.

(Continued)

```
---
Signif. codes:  0 '***' 0.001 '**' 0.01 '*' 0.05 '.' 0.1 ' ' 1

(Dispersion parameter for poisson family taken to be 1)

    Null deviance: 261.15  on 59  degrees of freedom
Residual deviance: 248.95  on 57  degrees of freedom
AIC: 493.02

Number of Fisher Scoring iterations: 5

> summary(overpoisson)

Call:
glm(formula = Carry ~ Length + Age, family = quasipoisson("log"),
  data = df1)

Deviance Residuals:
    Min       1Q     Median       3Q       Max
-5.5467   -0.8425    0.2225    1.0914    2.4431

Coefficients:
                Estimate    Std. Error    t value    Pr(>|t|)
(Intercept)     1.977750     1.251924       1.580     0.1197
Length          0.049512     0.025952       1.908     0.0615
Age            -0.004969     0.007813      -0.636     0.5274
---
Signif. codes:  0 '***' 0.001 '**' 0.01 '*' 0.05 '.' 0.1 ' ' 1

(Dispersion parameter for quasipoisson family taken to be 2.919268)

    Null deviance: 261.15  on 59  degrees of freedom
Residual deviance: 248.95  on 57  degrees of freedom
AIC: NA

Number of Fisher Scoring iterations: 5

> summary(zeropoisson)

Call:
zeroinfl(formula = Carry ~ Length + Age, data = df1, dist = "poisson",
    link = "log", control = zeroinfl.control(start = list(count = c(-2,
      0.9, 0.1), zero = c(-2, -0.2, -0.5)))))

Pearson residuals:
    Min       1Q     Median       3Q       Max
-2.6227   -0.5237    0.1147    0.7224    1.6381

Count model coefficients (poisson with log link):
                Estimate    Std. Error    z value    Pr(>|z|)
(Intercept)     3.053970     0.751935       4.061    4.88e-05 ***
Length          0.036558     0.015144       2.414    0.0158 *
Age            -0.009620     0.004807      -2.001    0.0454 *

Zero-inflation model coefficients (binomial with log link):
                Estimate    Std. Error    z value    Pr(>|z|)
(Intercept)     4.27320      7.72462        0.553     0.580
Length         -0.10127      0.14010       -0.723     0.470
Age            -0.02919      0.04915       -0.594     0.553
---
Signif. codes:  0 '***' 0.001 '**' 0.01'*' 0.05 '.' 0.1 ' ' 1

Number of iterations in BFGS optimization: 45
Log-likelihood: -169.3 on 6 Df
>
```

FIGURE 8.10 (CONTINUED)
Zero-inflated Poisson R script with output.

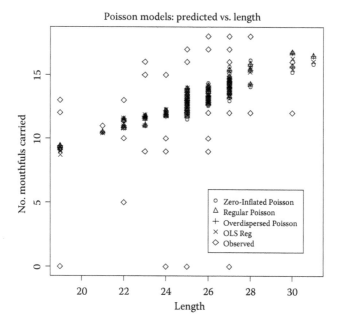

FIGURE 8.11
Predicted counts vs. Cichlid length.

Theoretical Aspects

Logistic Regression

The likelihood function for a sample of binary results, $y_1, y_2, ...,y_n$, and the associated regressor values $x_1, x_2,...,x_n$, is given by

$$L(\boldsymbol{\beta}) = \prod_{i=1}^{n} \pi^{y_i}(x_i)\left(1 - \pi(x_i)\right)^{1-y_i}$$

where

$$\pi(x_i) = \frac{e^{\beta_0+\beta_1 x_i}}{1+e^{\beta_0+\beta_1 x_i}} = \frac{1}{1+e^{\beta_0+\beta_1 x_i}} = \Pr\{Y = 1|x_i\}$$

The idea is to find values of β_0 and β_1 that maximize the likelihood function. Generally it is easier to maximize the logarithm of the likelihood function, since it involves sums and not products. Taking the natural log of $L(\boldsymbol{\beta})$, differentiating with respect to β_0 and β_1, and setting the partial derivatives equal to zero gives the equations:

$$\sum_{i=1}^{n}\left(y_i - \pi(x_i)\right) = 0$$

and

$$\sum_{i=1}^{n} x_i\left(y_i - \pi(x_i)\right) = 0$$

Inasmuch as both of these equations are nonlinear with respect to the two unknown parameters, there is no closed form solution. Rather, numerical approximation methods such as Newton–Raphson must be employed.

Once the model parameters are estimated, a measure of goodness is given by

$$D = -2\sum_{i=1}^{n}\left[y_i \ln\left(\frac{\hat{\pi}_i}{y_i}\right) + (1-y_i)\ln\left(\frac{1-\hat{\pi}_i}{1-y_i}\right)\right]$$

Where $\hat{\pi}_i = \dfrac{1}{1+e^{\hat{\beta}_0 + \hat{\beta}_1 x_i}}$ is the predicted probability that $Y = 1$ when $X = x_i$. If $y_i = 0$, then the first term in the sum is set to 0, and if $y_i = 1$, then the second term is set to 0. The quantity D is called the deviance (Hosmer and Lemeshow, 1989), and is analogous to the sums of squares for error in a usual multiple regression model. It can be used to compare the goodness of fit for models with different combinations of regressors. In general, lower deviance is more desirable.

It is fairly easy to generalize the logistic equation to multiple regressors:

$$P(X) = \Pr\{Y = 1|x\} = \frac{1}{1+e^{-\lambda(x)}} = \frac{1}{1+e^{-x'\beta}}$$

$$x' = \begin{bmatrix} 1 x_1 x_2 \cdots x_k \end{bmatrix}$$

is a vector of regressors (the "1" is the "regressor" for the intercept) and

$$\beta = \begin{bmatrix} \beta_0 \\ \vdots \\ \beta_k \end{bmatrix}$$

is a vector of unknown parameters.

Poisson Regression

The idea for Poisson regression, as in the case of logistic regression, is to simplify the process of estimating the values of the parameters. That is, if

$$\ln(\lambda) = \boldsymbol{\beta}^T x$$

so

$$\lambda = e^{\boldsymbol{\beta}^T x}$$

Then

$$\Pr\{Y = y_i | \lambda\} = \frac{\lambda^{y_i} e^{-\lambda}}{y_i!} = \frac{\left(e^{\boldsymbol{\beta}^T x}\right)^{y_i} e^{-\boldsymbol{\beta}^T x}}{y_i!}$$

The likelihood function is then

$$L(y_1, y_2, \ldots, y_n) = \prod_{i=1}^{n} \Pr\{Y = y_i | \lambda\} = \prod_{i=1}^{n} \frac{\left(e^{\boldsymbol{\beta}^T x}\right)^{y_i} e^{-\boldsymbol{\beta}^T x}}{y_i!}$$

It turns out to be easy (at least numerically) to find values of the β_j, $j = 0, \ldots, k$ (call these values $\hat{\beta}_j$, $j = 0, \ldots, k$) that maximize the log of the likelihood function for a given set of count data, $y_1, y_2, \ldots, y_n$.

Overdispersed Poisson

The general expression for the mass function of the negative binomial is

$$\Pr\{K = k | r, p\} = \binom{k + r - 1}{k} p^k (1 - p)^r$$

The parameter p is a probability, so it is a number between zero and one. The parameter r can be any real number, but it is usually nonnegative. The expected value and variance of K are

$$E[K | r, p] = \frac{rp}{1 - p}$$

$$V[K | r, p] = \frac{rp}{(1 - p)^2}$$

Since $0 < p < 1$, $V[K|r,p] > E[K|r,p]$. Thus, the negative binomial is like a Poisson that is overdispersed. An alternative to using the negative binomial is called the *quasipoisson* (Wedderburn, 1974). In this model, it is assumed that

$$E[K] = \lambda$$

and

$$V[K] = \theta\lambda = \theta E[K]$$

That is, the variance of the count variable K is assumed to be proportional; to the mean, λ. The parameter θ is called the *overdispersion parameter*.
The negative binomial could be thought of as a special case of the quasi-poisson, since

$$V[K] = \frac{rp}{(1-p)^2} = \frac{1}{(1-p)^2} E[K]$$

Zero-Inflated Poisson

If p represents the probability that an individual would always yield a zero, and if Y is the count variable, then (Johnson et al., 1992)

$$Pr\left\{Y_i = 0 \middle| \lambda\right\} = p + (1-p)\frac{(\lambda)^0 e^{-\beta^T x}}{0!}$$

In general,

$$Pr\left\{Y = y_i \middle| \lambda\right\} = \begin{cases} p + (1-p)\dfrac{(\lambda)^0 e^{-\beta^T x}}{0!}, & y_i = 0 \\[4mm] (1-p)\dfrac{(\lambda)^{y_i} e^{-\beta^T x}}{y_i!}, & y_i > 0 \end{cases}$$

The expected value of Y is

$$E\left[Y \middle| p, \lambda\right] = (1-p)\lambda$$

And its variance is

$$V\left[Y \middle| p, \lambda\right] = \lambda(1-p)(1+p\lambda)$$

Of course, as in the usual Poisson regression case, λ is defined in terms of the regressors:

$$\ln(\lambda) = \beta^T x$$

The variable Y therefore has a nonhomogeneous Poisson distribution, in that the probabilistic part has parameters depending on the values of the regressors.

Key Points for Chapter 8

- Odds are the ratios of probabilities; the probability of an "event" divided by the complementary probability.
- Binary responses can be functions of both continuous and discrete factors.
- A model using the logit transformation can be fit using binary response data.
- The coefficients of such models are actually the logarithms of odds.
- The model, called a *logistic regression*, gives predicted probabilities as a function of the regressors.
- Count-type data can be modeled through Poisson regression models.
- In some cases, count data require an additional overdispersion parameter.
- When more zero counts are observed than expected, a zero-inflated Poisson model may provide greater insight than the usual Poisson regression approach.

Exercises and Questions

1. In Example 8.4, both ordinary and overdispered Poisson regressions were performed. Were the conclusions of one more reliable than the other? Why?

2. Consider a situation where there is a continuously valued response variable, and multiple continuous regressors. Are there any scenarios where is would be advantageous to transform the continuous response into a binary variable (e.g., response values above a threshold are mapped to one; those below the threshold are mapped to zero)?

3. In the Cichlid cooperative behavior example, which of the models provided the best predictions? What criteria did you use to make your selection?

9

Multivariate Analyses: Dimension Reduction, Clustering, and Discrimination

General Ideas

Very often in behavioral ecology, multiple different variables are measured or observed on each individual or experimental unit in a study. Having multiple measurements like this often gives rise to the following questions:

1. Is there some way to reduce the dimensionality of the total number of variables into something more manageable?
2. Do individuals group or cluster based on all the different variables combined?
3. If the individuals group or cluster based on all the variables, do the groups correspond to some a priori ideas of how they might be grouped?

For example, imagine three related species of flowering plants that depend on animals for pollination. One of these species is pollinated by bees, another is pollinated by moths, and the third is pollinated by beetles. All of three species have flowers that produce a cocktail of volatile chemical compounds for the purpose of attracting pollinators. Since different pollinators are attracted to different scents, a researcher might want to know whether the chemical composition of the floral volatiles differs between the three plant species. If there was only one chemical analyte of interest, then ANOVA could be used to answer the question, "Do the species differ in their average analyte concentration?" A histogram of the analyte concentration might show multimodality, indicating the possibility that the plants have some natural clustering, or grouping, possibly, but not necessarily, related to their species. However, suppose that there are many different chemical analytes. Beyond two different chemical analytes, it would be very difficult to perceive any grouping from a histogram. Beyond three analytes, it becomes nearly impossible. The general methodology used to analyze multiple response variables simultaneously is called multivariate analysis.

Dimension Reduction: Principal Components

Usually the multiple variables measured are correlated in some fashion. Thus, some of the information from one variable may be obtained from another. In fact, the correlations of all pairs of variables can be obtained using the R function *cor(x=df)*, where df is the set of columns of variables. This is mostly just a description of the degree to which each pair of variables is related. One way to reduce the data set into a smaller number of variables, called *principal components analysis* (PCA), takes advantage of these correlations. It transforms the data into a set of statistically independent (uncorrelated) linear combinations of the original variables. These linear combinations are ranked in terms of their order of importance in describing the variability in the whole data set. The linear combinations are called principal components, and they are based on the eigenvectors of the correlation matrix as computed by the *cor()* function. The first principal component is the linear combination of the original variables that is associated with the greatest variability between all the individuals. The number of principal components is equal to the number of the original variables, but often after the first or second components, the remaining components are not as important. Thus, instead of having to consider whether differences in each original variable are significant, the analyst might only need to consider whether differences in the first and maybe the second principal component are significant.

To provide some intuition about principal components, consider a case with two variables, call them X1 and X2. Suppose that the correlation coefficient between X1 and X2 is approximately 0.7388. The plot in Figure 9.1 shows X2 plotted against X1, with two perpendicular lines. The first line (with the positive

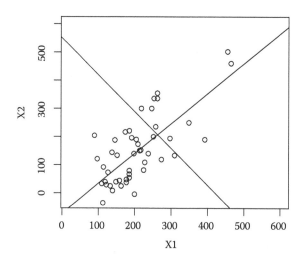

FIGURE 9.1
Two variables and their principal component axes.

slope) is actually the simple linear regression of X2 against X1. The second line is perpendicular to the regression line. The regression line is the first principal component axis; there is the greatest amount of variation alone this line. The second line is the second principal component, and corresponds to the axis of the second largest variation. There are exactly two principal components.

Clustering

The next multivariate question we will ask is whether there is any natural grouping, or clustering, of individuals. Cluster analysis is a collection of methods for finding natural groupings, or clusters, of individuals using multiple variables (Everitt et al., 2011). We will first discuss a class of cluster analysis called *k means*. In k means clustering, the analyst first specifies how many groups to look for, and then various algorithms are used to classify individuals into one of k groups. The disadvantage of k means clustering is that it requires the analyst to specify the number of groups a priori. This methodology may work best if there is at least a suspicion that individuals fall into a specific number of groups. It can be used to either confirm or refute the a priori suspicion. Another class of cluster analysis, which we will also present, is called hierarchical clustering, where the closeness or distance between individuals is used to create a hierarchy of groupings.

Discrimination

Discrimination is the process of sorting individuals, usually based on multiple variables, into predefined (a priori) categories or groups. This process is usually accompanied by a predictor formula that allows new individuals to be identified with one of the a priori categories. In this chapter, we will only present something called linear discriminant analysis (LDA) which will yield a set of linear classifying functions, $LD_i(x)$, for predicting the group or category to which an individual belongs.

The larger the absolute value of the entry, the more important the variable. If p is the number of variables measured or observed for each individual, and g is the number of possible groups to which an individual may belong, the number of linear discriminant functions required to categorize individuals is the lesser of p and $g-1$, where g is the number of groups.

Perhaps the simplest method for predicting the category or group to which an individual belongs assumes that the linear discriminant function values, $LD_1(x)$, $LD_2(x)$,...$LD_{g-1}(x)$ are jointly normally distributed. Since the LD_i are statistically independent, their joint probability distribution is easy to compute. The idea is to compute all the LD_i for each individual, then compute the means and standard deviations of the LD_i for each group or category. Then, using the normality assumption, the probability that an individual's set of LD_i, $i = 1$, $g-1$, come from each of the g groups can be computed as the joint likelihood of observing $LD_1(x)$, $LD_2(x)$,...,$LD_{g-1}(x)$, given that an individual belongs to each of the groups in turn. The analyst can then pick the group that maximizes this likelihood.

MANOVA

MANOVA is multivariate ANOVA; in other words, ANOVA in which there is more than one response variable. In MANOVA, the response is actually a matrix of variables, which may be correlated with each other. MANOVA is used to answer the question of whether the groups or categories are distinct, at least on the average, in terms of the multivariate space defined by all the response variables. In this sense, MANOVA is answering the same question as linear discriminant analysis, but it does not provide a predictive set of equations that can be used to classify new individuals. However, MANOVA can be used to discriminate for multiple categorizing factors simultaneously.

The advantage MANOVA has over PCA or LDA is that the entire dimensionality of the response variable space is incorporated, without having to select a single linear combination (e.g., PC1 or LD1) to rely upon for making a determination about the significance of differences between groups. The disadvantage of MANOVA is that it does not provide any way to determine which of the response variables is most important, nor does it provide any means of categorizing a new observation as does LDA.

Examples with R Code

Dimension Reduction: Principal Components

The vocalizations of a single species often differ in their acoustic structure across different parts of the species' range. When vocalizations can be grouped into discrete clusters based on their geographic origin, rather than varying from place to place along a graded continuum, the species is said to exhibit vocal dialects. Suppose that a researcher wants to determine whether dialects exist within a population of red-lored amazon parrots (*Amazona autumnalis*) in Ecuador. Parrots give loud contact calls to keep in touch with their flockmates, and the researcher suspects that there are three different dialects of these contact calls, corresponding to three adjacent geographical areas (let's call them s1, s2, and s3). She records contact calls from parrots in all three regions, and then measures the following acoustic features from each recorded call:

Q1Freq: The frequency below which 25 percent of the energy of the call is contained

BW90: The bandwidth within which 90 percent of the energy of the call is contained

CenterFreq: The frequency below which 50 percent of the energy of the call is contained

CenterTime: The time before which 50 percent of the energy of the call is contained

CTminusBT: The time from the start of the call until the CenterTime is reached

Dur90: The timespan during which 90 percent of the energy of the call is contained

IQR.Dur: The difference between the time before which 75 percent of the energy of the call is contained and the time before which 25 percent of the energy of the call is contained

PeakTime: The time at which the loudest point in the call occurs

PTminusBT: The time from the start of the call until the PeakTime is reached

These variable names are used by Cornell University's Raven Interactive Sound Analysis software (see http://www.birds.cornell.edu/brp/raven/RavenOverview.html).

As a first step toward deciding whether these putative dialects are in fact unique, the researcher may perform a principal components analysis (PCA) to reduce the dimensionality of the space. The first step is to extract the principal components from the raw acoustic measurements. Then, using the first principal component as the response variable in an ANOVA, decide whether or not the birds' scores along the first principal component axis differ by region. The code and associated output are shown in Figure 9.2.

Figure 9.3 shows the residuals against predicted (average) values for each Region. Figure 9.4 shows boxplots of the PC1 data by Region.

A summary (Table 9.1) of the *prcomp()* object, pcomp, shows the relative importance of the PCs. For these data, PC1 accounts for over 41 percent of the total variability. Furthermore, PC8 and PC9 account for almost nothing.

An ANOVA with PC2 as the response variable and subsequent Tukey HSD analysis shows that while there are significant differences in the PC2 axis between the Region levels, the difference between s1 and s3 does not appear to be significant. Thus, it may be true that while s1 and s3 are distinct, they are somewhat related (see Figure 9.5).

The last ANOVA we will do in this section is for PC3. Figure 9.6 shows the output.

For PC3, the only regions that appear to differ significantly are s2 and s3. Considering all three of these analyses together, we may hypothesize that regions s1, s2, and s3 are in fact unique. However, s1 and s3 are more closely related to each other than either is to s2. Furthermore, s1 and s2 are closer than are s3 and s2. Figure 9.7 shows a plot of the first two principal components, PC1 and PC2, with the region identified for each point.

```
setwd("C:\\Users\\SMFSBE\\Statistical Data & Programs")

#library(MASS)

df1 <-read.csv("20161227 Example 9.2 Multivariate Acoustic Spectral Data.csv")
#
# VARIABLES:
# Region
# Q1Freq
# BW90
# CenterFreq
# CenterTime
# CTminusBT
# Dur90
# IQR.Dur
# PeakTime
# PTminusBT
#
attach(df1)

#PRINCIPAL COMPONENTS ANALYSIS
#compute prcomp for the original data (i.e. the acoustic measurements of interest)

dfcomp <-
cbind(Q1Freq,BW90,CenterFreq,CenterTime,CTminusBT,Dur90,IQR.Dur,PeakTime,PTminusB
T)
pcomp <-prcomp(dfcomp,center=TRUE,scale=TRUE)


#Creating a new dataframe which is df1 plus the principle component scores
df2 <-cbind(df1,pcomp$x)

detach(df1)
attach(df2)

anovapc1 <-aov(PC1 ~ Region,na.action="na.omit")

summary(anovapc1)
TukeyHSD(anovapc1,"Region")

dev.new()
plot(fitted(anovapc1),resid(anovapc1))

meanpc1 < - tapply(PC1,Region,mean)

dev.new()

boxplot(PC1 ~ Region)


OUTPUT_____

> summary(anovapc1)

             Df     Sum Sq    Mean Sq    F value    Pr(>F)
Region        2     295.7     147.84     75.57      <2e-16 ***
Residuals   163     318.9       1.96
---
```

FIGURE 9.2
Principal components analysis of acoustic signal data.　　　　　　　　　*(Continued)*

```
Signif. codes:   0 '***' 0.001 '**' 0.01 '*' 0.05 '.' 0.1 ' ' 1

>

> dev.new()

> TukeyHSD(anovapc1,"Region")

  Tukey multiple comparisons of means

    95% family-wise confidence level


Fit: aov(formula = PC1 ~ Region, na.action = "na.omit")


$Region
                diff           lwr          upr        p adj
s2-s1     0.7794772     0.1012089     1.457745    0.0197983

s3-s1    -2.1840582    -2.8522315    -1.515885    0.0000000

s3-s2    -2.9635354    -3.5511109    -2.375960    0.0000000


>

> dev.new()

> plot(fitted(anovapc1),resid(anovapc1))

>

> meanpc1 <-tapply(PC1,Region,mean)

> dev.new()

> boxplot(PC1 ~ Region)
```

FIGURE 9.2 (CONTINUED)
Principal components analysis of acoustic signal data.

Clustering

In the parrot dialect example, the three regions suggest an a priori clustering. Thus, we might begin searching for three clusters, corresponding to the three regions. Figure 9.8 shows the R code for performing the k means clustering with function *kmeans*(), together with an ANOVA on the k = 3 clusters using PC1 has the response.

The difference between clusters 1 and 2 is not significant at the 5-percent level, but the differences between clusters 3 and 1 and between clusters 3 and 2 are. This appears to be in concert with the conclusions that we drew based on the PCA and ANOVAs. Figure 9.9 shows the plot of PC1 against PC2, but with clusters identified instead of regions.

Another class of clustering methods is called *hierarchical clustering*. Hierarchical clustering attempts to provide a set of linkages between groups of individuals and within groups. It is particularly useful when the analyst is interested in how clusters are nested within one another; for example, when investigating phylogenetic relationships (Chakerian and Holmes, 2012) or

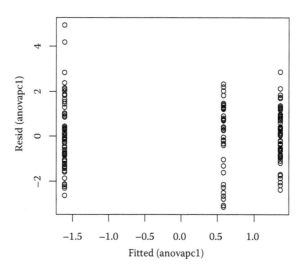

FIGURE 9.3
Residuals versus predicted: Region ANOVA.

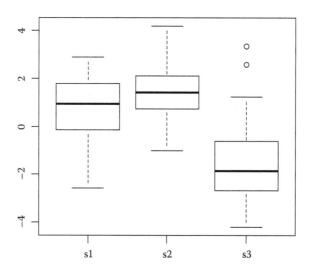

FIGURE 9.4
Boxplots of PC1 by Region.

the social structure of multilevel animal societies (Whitehead, 2008). There are many different hierarchical clustering algorithms. In general they are based on a measure of distance between individuals in a multivariate space. Perhaps the most well-known distance measure is Euclidean distance. If $X_1 = [x_{11}, x_{12},..., x_{1k}]$ is a vector of values for k variables on a particular individual, an $X_2 = [x_{21}, x_{22},..., x_{2k}]$ is the set of measurements for the same variables, but

TABLE 9.1

Summary of the PCA

	PC1	PC2	PC3	PC4	PC5	PC6	PC7	PC8	PC9
	\> **Summary(pcomp)**								
	Importance of Components:								
Standard deviation	1.9299	1.3255	1.0702	0.9551	0.9419	0.5374	0.3978	0.3562	0.0000
Proportion of variance	0.4138	0.1952	0.1273	0.1014	0.0986	0.0321	0.0176	0.0141	0.0000
Cumulative proportion	0.4138	0.6091	0.7363	0.8377	0.9362	0.9683	0.9859	1.0000	1.0000

```
>
> anovapc2 <-aov(PC2 ~ Region,na.action="na.omit")
>
> summary(anovapc2)
              Df       Sum Sq      Mean Sq     F value     Pr(>F)
Region         2        39.46       19.730      12.84       6.62e-06 ***
Residuals    163       250.45        1.536
---
Signif. codes:  0 '***' 0.001 '**' 0.01 '*' 0.05 '.' 0.1 ' ' 1
> TukeyHSD(anovapc2,"Region")
  Tukey multiple comparisons of means
  95% family-wise confidence level

Fit: aov(formula = PC2 ~ Region, na.action = "na.omit")

$Region
            diff          lwr          upr        p adj
s2-s1   1.0807857    0.4796712    1.6819002    0.0001048
s3-s1   0.1174681   -0.4746997    0.7096359    0.8858292
s3-s2  -0.9633176   -1.4840557   -0.4425795    0.0000638
```

FIGURE 9.5
Output for ANOVA with PC2.

on a different individual, then the Euclidean distance between the two individuals is given by

$$d(X_1, X_2) = \sqrt{\sum_{i=1}^{k}(x_{1i} - x_{2i})^2}$$

The R function *dist()* takes a matrix or dataframe with just the multivariate measurements or observationsand computes a distance matrix for all the individuals, represented by the rows of the input. The default distance method is

```
>
> anovapc3 <-aov(PC3 ~ Region,na.action="na.omit")
> summary(anovapc3)
               Df       Sum Sq      Mean Sq     F value      Pr(>F)
Region          2         9.07        4.535       4.109      0.0182 *
Residuals     163       179.91        1.104
---
Signif. codes:  0 '***' 0.001 '**' 0.01 '*' 0.05 '.' 0.1 ' ' 1
> #interaction.plot(x.factor=factor(GroupNo),trace.factor=Region,response=PC1)
>
>
> TukeyHSD(anovapc3,"Region")
  Tukey multiple comparisons of means
    95% family-wise confidence level

Fit: aov(formula = PC3 ~ Region, na.action = "na.omit")

$Region
             diff           lwr          upr         p adj
s2-s1  -0.2548351   -0.76431707    0.2546469     0.4649399
s3-s1   0.2796134   -0.22228573    0.7815125     0.3873683
s3-s2   0.5344485    0.09309046    0.9758065     0.0130644
```

FIGURE 9.6
Output for ANOVA with PC3.

Euclidean. This in turn can be used in the function *hclust()*, which will perform the hierarchical clustering. The *hclust()* function uses a distance matrix,
which can be created using the *dist()* function, to produce an object that can be
plotted as a dendrogram. The dendrogram is so-called because it resembles a
tree with branches that connect groups of individuals based on the distance
measure, and should look instantly familiar to any biologist who is used to

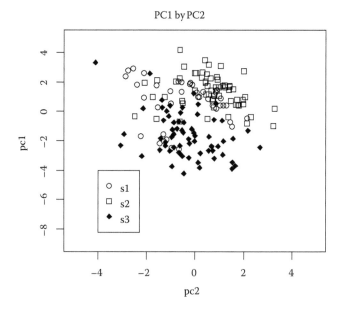

FIGURE 9.7
Plot of first two principal components with region.

```
setwd("C:\\Users\\SMFSBE\\Statistical Data & Programs")

#library(MASS)

df1 <-read.csv("20161227 Example 9.2 Multivariate Acoustic Spectral Data.csv")
#
# VARIABLES:
# Region
# Q1Freq
# BW90
# CenterFreq
# CenterTime
# CTminusBT
# Dur90
# IQR.Dur
# PeakTime
# PTminusBT
#
attach(df1)

#K-MEANS CLUSTER ANALYSIS
#Creating a dataframe of just the response variables


dfcluster <-
cbind(Q1Freq,BW90,CenterFreq,CenterTime,CTminusBT,Dur90,IQR.Dur,PeakTime,PTminusB
T)

#Conducting a k-means clustering algorithm on the dataframe dfcluster, using 3 clusters and the
Hartigan-Wong algorithm
clus1 <-kmeans(dfcluster,centers=3,algorithm=c("Hartigan-Wong"))

#Renaming the column with cluster ID from "clus1" to "clusid"
clusid <-clus1$cluster
fclusid <-as.factor(clusid)

#PRINCIPLE COMPONENTS ANALYSIS
#compute princomp for the original data (i.e. the acoustic measurements of interest)

pcomp <-prcomp(dfcluster,center=TRUE,scale=TRUE)
#Creating a new dataframe which is df1 plus the cluster ID column (clusid) and principal
#components
df2 <-cbind(df1,clusid,fclusid,pcomp$x)
detach(df1)

    attach(df2)
    #
    # Test to see if clusters appear to differ significantly
    #
    anova(lm(PC1 ~ fclusid))

    #create plot points
    pcomp11 <-PC1[which(clusid==1)]
    pcomp12 <-PC2[which(clusid==1)]
    pcomp21 <-PC1[which(clusid==2)]
    pcomp22 <-PC2[which(clusid==2)]
    pcomp31 <-PC1[which(clusid==3)]
    pcomp32 <-PC2[which(clusid==3)]

    plot(pcomp12,pcomp11,pch=21,xlim=c(-5,5),ylim=c(-9,5),xlab="pc2",ylab ="pc1",main= "PC1
    by PC2")

    # Now add the points from the other clusters
    points(pcomp22,pcomp21,pch=22)
    points(pcomp32,pcomp31,pch=18)

    # Create the legend; Note that pch in legend is specified in the order of group names:

    legend(-4,-4,c("Cluster1","Cluster2","Cluster3"),pch=c(21,22,18))
```

FIGURE 9.8
K means clustering with ANOVA. *(Continued)*

```
OUTPUT

> #
> # Test to see if clusters appear to differ significantly
> #
> anova(lm(PC1 ~ fclusid))
Analysis of Variance Table

Response:PC1
             Df     Sum Sq    Mean Sq    F value      Pr(>F)
fclusid       2     355.29     177.65     111.69    < 2.2e-16 ***
Residuals   163     259.25       1.59
---
Signif. codes:  0 '***' 0.001 '**' 0.01 '*' 0.05 '.' 0.1 ' ' 1

> TukeyHSD(aov(PC1 ~ fclusid),"fclusid")
  Tukey multiple comparisons of means
    95% family-wise confidence level

Fit: aov(formula = PC1 ~ fclusid)

$fclusid
          diff          lwr         upr        p adj
2-1   0.5002748   -0.1215998    1.122149   0.1411293
3-1  -2.9045029   -3.4271817   -2.381824   0.0000000
3-2  -3.4047777   -4.0572771   -2.752278   0.0000000
```

FIGURE 9.8 (CONTINUED)
K means clustering with ANOVA.

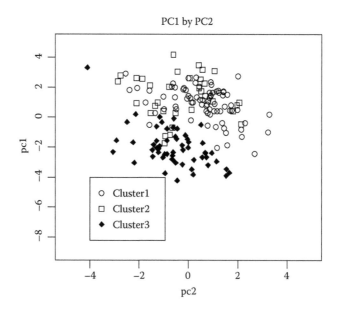

FIGURE 9.9
PC1 against PC2 with clusters identified.

```
dev.new()
hier <-hclust(dist(dfcluster),method="complete")
plot(hier)
```

FIGURE 9.10
Hierarchical clustering code.

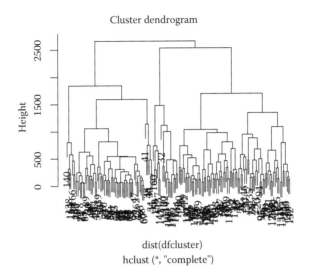

FIGURE 9.11
Dendrogram.

reading phylogenetic trees. In fact, hierarchical clustering is one of the methods through which phylogenetic trees may be inferred, and the phylogenetic tree is itself a type of dendrogram. Figure 9.10 shows the code used to make such a dendrogram, and Figure 9.11 shows the dendrogram for the acoustic data.

A chi-squared test can be used to assess whether there is a relationship between clusters and regions. Figure 9.12 shows the results of a permutation chi-squared test.

While the relationship between cluster and region does not appear to be random ($p = 0.0004998$), the s2 region seem to be split between clusters 1 and 3, and the majority of region s1 data are associated with cluster 3. Thus there is some question as to whether the calls can be discriminated in terms of the three regions.

```
> chisq.test(Region,fclusid,correct=False,simulate.p.value=TRUE,B=2000)

    Pearson's Chi-squared test with simulated p-value (based on 2000
    replicates)

data:  Region and fclusid
X-squared = 71.557, df = NA, p-value = 0.0004998

> table(Region,fclusid)
            fclusid
Region 1 2 3
     s1    6  7 26
     s2   18  3 40
     s3    9 47 10
```

FIGURE 9.12
Relationship between clusters and regions.

Discrimination

In the parrot dialect example, there are $g = 3$ groups or categories in which individual calls may be classified: s1, s2, and s3. The R function *lda*(), in library MASS, will perform the linear discriminant analysis. The function *predict*() will take an *lda*() object and compute the predicted group for each individual in the set.

The R code in Figure 9.13 shows how the linear discriminant analysis is executed, and the output of a table showing the numbers of individuals of each region (variable Region) and the category in which the predict function placed them (ldaclass). Figure 9.14 shows a plot of the LD1 and LD2 values, with the actual region for each individual.

Even though the linear discriminant functions were calculated using the data that are plotted, the table shows that the classification is not perfect. However, it seems that the linear discriminant functions, in this case, did

```
#LINEAR DISCRIMINANT ANALYSIS
#the ldafunction is contained in the package "MASS", so I have to load this package first
#Conduct a linear discriminant analysis, using "Region" as the grouping variable
#disc1 is an lda object that I am creating
#CV=TRUE means I am asking R to give me the results for leave-one-out cross-validation
disc1 <-lda(Region ~ Q1Freq + BW90 + CenterFreq + CenterTime + CTminusBT + Dur90 +
IQR.Dur + PeakTime + PTminusBT, data=df2, na.action="na.omit", CV=FALSE)
dev.new()

#creating a column that has the group classifications as predicted by lda
pred1 <-predict(disc1,newdata=df2)

#renaming this column "ldaclass"
ldaclass <-pred1$class
LD1 <-pred1$x[,1]
LD2 <-pred1$x[,2]

#creating a new dataframe that has the original data plus ldaclass
df3 <-cbind(df2,ldaclass,LD1,LD2)
detach(df2)
attach(df3)
#creating a table displaying "Region" and "ldaclass"
table(Region,ldaclass)

ld1s1 <-LD1[which (Region=="s1")]
ld2s1 <-LD2[which (Region=="s1")]
ld1s2 <-LD1[which (Region=="s2")]
ld2s2 <-LD2[which (Region=="s2")]
ld1s3 <-LD1[which (Region=="s3")]
ld2s3 <-LD2[which (Region=="s3")]

plot(ld2s1,ld1s1,main="Biplot of Linear Discriminant Function Values",pch=1,xlim=c(
4,4),ylim=c(-4,4),xlab="LD2",ylab="LD1")
points(ld2s2,ld1s2,pch=2)
points(ld2s3,ld1s3,pch=3)
OUTPUT
> table(Region,ldaclass)
          ldaclass
Region    s1  s2  s3
    s1    25   7   7
    s2     4  54   3
    s3     6   4  56
```

FIGURE 9.13
Linear discriminant analysis code.

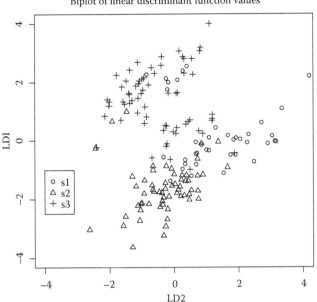

FIGURE 9.14
Plot of LD1 by LD2, with region identified.

```
> table(ldaclass,clusid)
         clusid
ldaclass    1    2    3
      s1    0   27    8
      s2    0   46   19
      s3   56    6    4
```

FIGURE 9.15
LDA classification by cluster ID.

a better job in classifying individuals than did the clustering. Figure 9.15 shows the tabulation of results classified by *lda()* against the clusters.

It is apparent that *lda()* had greater success in classifying individuals than the cluster analysis. Of course, the a priori knowledge of which region each call belonged to was not used in the cluster analysis.

MANOVA

We performed a MANOVA for the acoustic data. The code and output are shown in Figure 9.16. The variables CenterTime and PeakTime were almost identical in value, so that only one of them could be included in the MANOVA.

```
response <-
cbind(Q1Freq,BW90,CenterFreq,CenterTime,CTminusBT,Dur90,IQR.Dur,PTminusBT)
man <-manova(data=df2,response ~ Region,na.action="na.omit")
summary(man)
```

OUTPUT

```
> summary(man)
           Df     Pillai        approx F      num Df      den DfPr(>F)
Region      2    0.98779         19.151           16         314        < 2.2e-16 ***
Residuals  163
---
Signif. codes:  0 '***' 0.001 '**' 0.01 '*' 0.05 '.' 0.1 ' ' 1
```

FIGURE 9.16
MANOVA code and output.

```
> man <-manova(data=df2,response ~ Region+ factor(GroupNo),na.action="na.omit")
> summary(man,test="Wilks")
                   Df    Wilks      approx F  num Df   den Df    Pr(>F)
Region              2   0.044724     70.377       16   302.00   <2.2e-16 ***
factor(GroupNo)     5   0.045321     17.080       40   660.99   <2.2e-16 ***
Residuals         158
---
Signif. codes:  0 '***' 0.001 '**' 0.01 '*' 0.05 '.' 0.1 ' ' 1
```

FIGURE 9.17
Two-factor model for MANOVA.

Another important feature of MANOVA is that, like its univariate counter-part (ANOVA), multiple classification or grouping factors can be considered simultaneously. For example, suppose that in addition to coming from one of three different geographical regions, the red-lored amazon parrots can also be classified according to which social group they belong.

The code and output shown in Figure 9.17 shows how both factors can be considered in the same model simultaneously.

Theoretical Aspects

We will briefly introduce a few concepts related to multivariate normal distributions. Two random variables, X_1 and X_2, are said to have a joint or multi-variate normal distribution if their joint density function can be expressed as

$$f(x) = \frac{1}{\sqrt{2\pi}^k |\Sigma|^{\frac{1}{2}}} \exp\left(-\frac{1}{2}(x-\mu)^T \Sigma^{-1}(x-\mu)\right)$$

where

$$k = 2, \ x = \begin{bmatrix} x_1 \\ x_2 \end{bmatrix}, \ \mu = \begin{bmatrix} \mu_1 \\ \mu_2 \end{bmatrix}, \text{ and } \Sigma = \begin{bmatrix} \sigma_1^2 & \sigma_{12} \\ \sigma_{21} & \sigma_2^2 \end{bmatrix}$$

That is, x is a vector of values for X_1 and X_2, μ is a vector of expected values for the two random variables, and Σ is the covariance matrix. Of course, it is possible for $k > 2$ random variables to have a joint distribution. While multivariate normality may or may not be rare in nature, it is a point of departure for the methods discussed in this chapter.

Principal Components

As mentioned earlier, principal components are a set of alternate axes in a multidimensional space defined by all the variables measured on each individual. The new axes are ordered in terms of the amount of total variation in the sample accounted for by each new axis. The axes are determined by the eigenvectors of the correlation matrix of the original variables. Sometimes the covariance matrix is used, although using the covariance matrix creates some difficulties in terms of scale, since covariances have units that are products of the units of the two variables that co-vary. Thus, we typically rely on the correlation matrix. Often, before the correlation matrix and its eigenvectors are computed, each variable is shifted and scaled, by subtracting their means and dividing by their standard deviations. The eigenvector values represent coefficients for each of the original variables regardless of whether or not those variables were shifted and scaled. Thus, if

$$v_j = \begin{bmatrix} v_{1j} \\ \vdots \\ v_{pj} \end{bmatrix}$$

Represents the jth eigenvector of the correlation matrix, then if the original variables are Y1, Y2, ..., Yp, the jth principal component would be

$$PC_j = v_{1j}Y_1 + v_{2j}Y_2 + \ldots + v_{pj}Y_p$$

So given a set of values of the original variables for an individual, say y_{i1}, y_{i2}, ..., y_{ip}, the individual's value for PCj would be

$$pc_{ij} = v_{1j}y_{i1} + v_{2j}y_{i2} + \ldots + v_{pj}y_{ip}$$

Each PC is a linear combination of the orginal variables. The PCs are ordered in terms of their percent of total variation that they account for (and thus importance) in the sample of individuals.

Discrimination

When there are several categories or groups of individual experimental units known to exist a priori, it is sometimes desirable to determine whether or not an individual can be classified into the correct category by having the values of certain measurements or observed variables for that individual. The first step is to collect the variables of interest on a sample of individuals. These individuals will be referred to as the *training set*. They will be used to empirically derive some equations that will allow new individuals to be classified. It is presumed that for the training set, the category or group for each individual in that set is known without error.

The following method is described by Anderson (1958). Suppose there are g groups, p variables for each individual, and n_k individuals in group k. Then let $\bar{x}_j^{(k)}$ represent the average of variable j ($j = 1, p$) for group k ($k = 1, g$). The vector of these averages for group k is the called mean vector for group k. Let $\bar{x}^{(k)}$ represent the mean vector for group k. If S represents the sample covariance matrix, and x_i represents the vector of observations of the p variables for individual i, then form the following functions:

$$f_{kl}(x_i) = \left[x_i - \frac{1}{2}\left(\bar{x}^{(k)} + \bar{x}^{(l)}\right) \right]^T S^{-1}\left[\bar{x}^{(k)} - \bar{x}^{(l)}\right]$$

for $k \neq l$. Then define:

$$f_{lk}(x_i) = -f_{kl}(x_i)$$

Then for $k = 1, g$, the grouping rules are to assign individual i to group k if

$$f_{kl}(x_i) \geq 0, \text{ for } l \neq k \text{ for } k = 1, g$$

The functions $f_{kl}(x_i)$ are called *linear discriminant functions*, since they are linear combinations of the vector x_i.

Another methodology is described by Armitage (1971). It involves partitioning the sums of squares matrix (i.e., the covariances without dividing by the sample sizes) into two additive pieces: Between groups and within groups. Let

x_{ijk} = the observation/value of the jth variable for the ith individual in group k

$$x_{i.k} = \begin{bmatrix} x_{i1k} \\ \vdots \\ x_{ipk} \end{bmatrix} = \text{vector of } p \text{ variable values for individual } i \text{ in group } k$$

$$\bar{x}^{(k)} = \begin{bmatrix} \bar{x}_{.1k} \\ \vdots \\ \bar{x}_{.pk} \end{bmatrix} = \text{vector of } p \text{ means for group } k$$

$$\bar{x} = \begin{bmatrix} \bar{x}_{.1.} \\ \vdots \\ \bar{x}_{.p.} \end{bmatrix} = \text{vector of } p \text{ means (one for each variable)}$$

averaged over groups and individuals

$$W_{ik} = \left[x_{i.k} - \bar{x}^{(k)} \right]\left[x_{i.k} - \bar{x}^{(k)} \right]^T = p \times p$$

matrix of products of within-group differences

$$B_k = \left[\bar{x}^{(k)} - \bar{x} \right]\left[\bar{x}^{(k)} - \bar{x} \right]^T = p \times p \text{ matrix of products of differences}$$

between kth group mean vector and overall mean vector

$$W = \sum_{k=1}^{g} \sum_{i=1}^{n_k} W_{ik} = p \times p \text{ within groups sums-of-squares matrix}$$

$$B = \sum_{k=1}^{g} B_k = p \times p \text{ between groups sums-of-squares matrix}$$

The total $p \times p$ sums-of-squares matrix, T, is the sum

$$T = W + B$$

The linear discriminant functions are the eigenvectors of the $p \times p$ matrix

$$W^{-1}B$$

Generally, these eigenvectors are first "normalized" by multiplying them by the reciprocal of their length, so that the normalized eignevectors all have length 1.

Like principal components, the p original variables are mapped into p linear combinations, where the combinations are independent of each other. So, if

$$v_1 = \begin{bmatrix} v_1 \\ \vdots \\ v_p \end{bmatrix}$$

is the normalized eigenvector corresponding to the largest eigenvalue of $W^{-1}B$ then the first linear discriminant function, LD_1, would be

$$LD_1(x) = v_1^T x$$

and x is a vector of the p variable values for an individual. The entries in each of the vectors v_i, i = 1, g, indicates the relative importance of each response variable.

MANOVA

Recall that the total sums-of-squares matrix can be expressed as a matrix sum:

$$T = B + W$$

If there are g groups or categories, and n_k observations per category, with p response variables, then let

$$V_w = p * (g - 1)$$

$$V_b = \sum_{i=1}^{g} n_i - g + 1$$

The null hypothesis is that the mean vector for the response variables is the same for all groups, and that, as in the case of LDA, the data come from a multivariate normal distribution. The R function *manova()* computes an "approximate" F statistic, by using a statistic called Wilk's Λ (Anderson, 1958). The Λ statistic is

$$\Lambda = \frac{|W|}{|T|} = \prod_{i=1}^{p} \frac{1}{1 + \lambda_i}$$

where λ_i are the eigenvalues of the matrix $W^{-1}B$. The Wilk statistic has something called an *incomplete beta distribution* (Anderson, 1958), which, computationally speaking, is generally not easy to use. Thus, a statistic whose distribution is approximately F is perhaps more desirable.

The approximate F statistic is

$$F = \frac{1 - \Lambda^{\frac{1}{t}}}{\Lambda^{\frac{1}{t}}} \frac{df_2}{df_1}$$

and

$$t = \sqrt{\frac{p^2 + (g-1)^2 - 4}{p^2 + (g-1)^2 - 5}}$$

$$df_1 = p * v_b$$

$$df_2 = \left(v_w + v_b - 0.5 * (p + v_b + 1)\right) * t - 0.5 * (p * v_b - 2)$$

The approximate F would be compared to the $(1-\alpha)100$ percentile of an F distribution with df_1 numerator degrees of freedom and df_2 denominator degrees of freedom.

Key Points for Chapter 9

- One objective of multivariate analysis is to reduce the dimensionality of the data, generally by finding strategic linear combinations of the variables.
- Individuals may "cluster" into naturally occurring groups that were not known to exist a priori.
- When categories or groupings exist a priori, techniques such as linear discriminant analysis can provide a means of classifying new individuals into groups.
- Whether the categories are identified a priori or by analysis, the actual separation between those categories can be tested.

Exercises and Questions

1. Discuss the differences between k means and hierarchical clustering. What arguments can be made for both methods?
2. Compare conclusions about the Example 9.2 data using principle components, linear discriminants, and MANOVA.

10

Bayesian and Frequentist Philosophies

General Ideas

Bayes' Theorem: Not Controversial

Bayes' theorem was described in Chapter 1. This is a simple example of how it might work. Suppose there are three subpopulations of an organism, $P1$, $P2$, and $P3$. Suppose further that the total population, $P1 + P2 + P3$, is partitioned in the following way:

$P1$: 20 percent
$P2$: 45 percent
$P3$: 35 percent

Consider a behavior, B. In $P1$, it is known (somehow) that 10 percent of the individuals in this subpopulation exhibit behavior B as a response to a particular stimulus, S. In $P2$, it is 25 percent, and in $P3$ it is 15 percent. Thus, we might formulate the following probability statements:

$Pr\{P1\} = 0.20$
$Pr\{P2\} = 0.45$
$Pr\{P3\} = 0.35$
$Pr\{B|P1\} = 0.25$
$Pr\{B|P2\} = 0.05$
$Pr\{B|P3\} = 0.15$

So, Bayes' theorem gives us a way of computing the probability that a randomly selected individual, randomly selected without regard to subpopulation, would respond to the stimulus with behavior B:

$$Pr\{B\} = Pr\{B|P1\}Pr\{P1\} + Pr\{B|P2\}Pr\{P2\} + Pr\{B|P3\}Pr\{P3\}$$
$$= (0.25)(0.20) + (0.05)(0.45) + (0.15)(0.35)$$
$$= 0.125$$

Thus, there is a 12.5-percent chance that a randomly selected individual would respond to the stimulus with behavior B. This is incontrovertible, and is the primary consequence of Bayes' theorem.

The next part of Bayes' theorem is what was once referred to as "inverse probability" (Jeffreys, 1961). Suppose we had an individual who responded to stimulus S with behavior B. Perhaps we would like to know from which subpopulation this individual came. We can use Bayes' theorem to compute the probabilities:

$$Pr\{P1|B\} = \frac{Pr\{B|P1\}Pr\{P1\}}{Pr\{B\}} = \frac{(0.25)(0.20)}{0.125} = 0.4000$$

$$Pr\{P2|B\} = \frac{Pr\{B|P2\}Pr\{P2\}}{Pr\{B\}} = \frac{(0.05)(0.45)}{0.125} = 0.1800$$

$$Pr\{P3|B\} = \frac{Pr\{B|P3\}Pr\{P3\}}{Pr\{B\}} = \frac{(0.15)(0.35)}{0.125} \approx 0.4200$$

One way to guess which subpopulation the individual comes from is to pick whichever option has the highest probability above (i.e., $P3$). The inverse probabilities are now generally referred to as "posterior" probabilities. The probabilities $Pr\{P1\}$, $Pr\{P2\}$, and $Pr\{P3\}$ are referred to as "prior" probabilities. The rule for making the choice based on maximizing the posterior probability is called a *Bayes' rule* (Gelman et al., 2000). That seems reasonable. However, here is where the controversy begins. The classic, or frequentist, philosophy is to not incorporate prior probabilities into the decision-making process. Rather, the frequentist would choose the alternative that maximizes the conditional probabilities $Pr\{B|P1\}$, $Pr\{B|P2\}$, and $Pr\{B|P3\}$, which would lead one to choose $P1$, and not $P3$. The probabilities concerning the data given the state of truth are called *likelihoods*. Thus, the frequentist seeks to maximize the likelihood function (a function over all the possibilities for the truth) whereas the Bayesian seeks to maximize the probability over the posterior distribution of the possibilities for the truth.

This simple example demonstrates the fundamental objectives of Bayesian and frequentist analyses. The major problem with Bayesian analyses is choosing the prior distribution; the major problem with frequentist methods is that there is no obvious way to update knowledge in the face of new data.

In many cases, the prior distribution and likelihood functions have parametric forms; that is, in addition to the variables of interest, the prior distribution has some parameters whose values are "updated" using the likelihood function. Generally, the variable of interest is a parameter, θ, whose value is unknown. However, there is some a priori idea of the possible range of values and frequencies within which this parameter might fall. The prior information is expressed as a density (continuous) or mass (discrete) function. The prior density in turn usually has parameters whose values will be updated once data are obtained. So, if θ is the parameter of interest, suppose it has a prior density of the form:

$$g'(\theta|\alpha')$$

The parameter α' is referred to as a *hyperparameter*, whose value describes the shape of the density for θ. Suppose x represents the data, and $f(x|\theta)$ represents the likelihood function for getting data x given the parameter θ. The prior density for θ will be updated using Bayes' theorem.

$$g''(\theta|\alpha'') = \frac{g'(\theta|\alpha')f(x|\theta)}{\int_{-\infty}^{+\infty} f(x|\xi)g'(\xi\alpha')d\xi}$$

Thus the prior density, g', is updated using the likelihood value for the data x. The posterior density for θ, g'', can be used as the new prior density if more data become available.

Conjugacy

There are certain pairs of distributional forms for priors and likelihoods that allow for a relatively simple, closed form means of deriving the posterior distribution. Furthermore, only some combinations of priors and likelihoods would yield a posterior distribution of the same family as the prior. Such combinations are called *conjugate pairs* (Gelman et al., 2000). In general, for conjugate pairs, Bayes' formula produces equations for updating the hyperparameters, which in turn define the posterior distribution.

There are several conjugate pairs that are useful to know.

Beta, Binomial

If Y is a binomial random variable with parameter p, and sample size n, then a conjugate prior for its likelihood function is the beta distribution. The likelihood function would be

$$L[y \mid p] = \binom{n}{y} p^y (1-p)^{n-y}$$

The beta prior density has two hyperparameters, α', β', and is

$$f\left(p \mid \alpha', \beta'\right) = \frac{\Gamma\left(\alpha' + \beta'\right)}{\Gamma(\alpha')\Gamma(\beta')} p^{\alpha'-1}(1-p)^{\beta'-1}$$

The posterior distribution of p would also be beta, with posterior hyperparameters

$$\alpha'' = \alpha' + y$$

$$\beta'' = \beta' + n - y$$

The mean (expected value) of a beta variable with parameters α and β is

$$E\left[p \mid \alpha, \beta\right] = \frac{\alpha}{\alpha + \beta}$$

This can sometimes be helpful in defining a prior beta distribution.

Poisson, Gamma

If Y is a Poisson random variable with parameter λ, then its likelihood function would be

$$L\left[y_1, y_2, \ldots, y_n \mid \lambda\right] = \prod_{i=1}^{n} \frac{\lambda^{y_i} e^{-\lambda}}{y_i!}$$

The prior gamma distribution with hyperparameters α' and β' would be

$$f\left(\lambda \mid \alpha', \beta'\right) = \frac{\beta'^{\alpha'}}{\Gamma(\alpha')} \lambda^{\alpha'-1} e^{-\beta'\lambda}$$

The posterior distribution of λ is also gamma, with hyperparameters:

$$\alpha'' = \alpha' + \sum_{i=1}^{n} y_i$$

$$\beta'' = \beta' + n$$

Note the similarity to the exponential, gamma conjugate pair. Recall that if Y has a Poisson distribution with parameter $\lambda \tau$, and T has an exponential distribution with parameter λ, then

$$Pr\{Y = 0|\lambda\tau\} = \frac{(\lambda\tau)^0 e^{-\lambda\tau}}{0!} = e^{-\lambda\tau} = Pr\{T \geq \tau|\lambda\}$$

Normal, Normal

Suppose Y has a normal distribution, where somehow the parameter σ is known. The likelihood function for a sample of n values is

$$L[y_1, y_2, \ldots, y_n] = \prod_{i=1}^{n} \frac{1}{\sqrt{2\pi}\sigma} \exp\left(-\frac{1}{2}\left(\frac{y_i - \mu}{\sigma}\right)^2\right)$$

The prior for μ with hyperparameters ξ' and τ' is

$$f(\mu|\xi', \tau') = \frac{1}{\sqrt{2\pi}\tau'} \exp\left(-\frac{1}{2}\left(\frac{\mu - \xi'}{\tau'}\right)^2\right)$$

The posterior is also normal, with hyperparameters

$$\xi'' = \frac{\frac{1}{\tau'^2}\xi' + \frac{n}{\sigma^2}\bar{y}}{\frac{1}{\tau'^2} + \frac{n}{\sigma^2}}$$

and

$$\frac{1}{\tau''^2} = \frac{1}{\tau'^2} + \frac{n}{\sigma^2}$$

If σ is not known, and in fact has its own prior distribution, the situation can become fairly complicated.

Monte Carlo Markov Chain (MCMC) Method

Sometimes the prior density and likelihood functions are not conjugate. In theory, a posterior density can be computed numerically. That is, for a given set of observed data, x, to get the posterior density $g''(\theta|\alpha'')$ we could first numerically evaluate the integral:

$$h(x) = \int_{-\infty}^{+\infty} f(x|\xi) g'(\xi|\alpha') d\xi$$

and then compute:

$$g''(\theta|\alpha'') = \frac{g'(\theta|\alpha') f(x|\theta)}{h(x)}$$

This is not too difficult if θ is a (scalar) parameter. If, however, θ is actually a vector of parameters, then the problem becomes more complex. Fortunately, a set of numerical procedures known as Monte Carlo Markov Chain (MCMC) methods provide a fairly convenient way to find a posterior distribution, even if the parameter θ is a vector. One helpful feature of MCMC is that the function $h(x)$ is never explicitly evaluated. The Monte Carlo part refers to using random numbers to evaluate integrals (Law and Kelton, 2015). The Markov Chain part refers to a probabilistic progression or process where the probability that the next value in the sequence depends entirely on the current value (Grimmett and Stirzaker, 2004). This is how the MCMC process works:

1   Initially and arbitrarily chose a value for θ, call it θ_0.

2. Randomly choose another value for θ, call it θ_1. You can use any method you like for randomly generating θ_1. Some researchers randomly choose a value from a normal distribution with mean θ_0 and some constant variance.

3. Compute the ratio

$$R = \frac{g'(\theta_1|\alpha') f(x|\theta_1)}{g'(\theta_0|\alpha') f(x|\theta_0)}$$

4. Randomly generate a number between zero and one, and call it u. If $R \geq u$, then the new value for θ is θ_1. Store it in the list of θs. Otherwise, store θ_0.

5. Assign to θ_0 the newly stored value.

6. Repeat the process n times from step (2).

When you have done this n times, you have a list of values for θ that would simulate a sample coming from the posterior density, given the data x.

This clever process is referred to as the Metropolis–Hastings algorithm (Gelman et al., 2000). The random generation is the Monte Carlo part. The Markov Chain part is due to the fact that the current value of the stored θ only depends on the immediately preceding value of θ_0, and not the whole history of previous θ_0 values.

Examples with R Code

Exponential, Gamma

Parasitoid wasps typically lay their eggs inside of a living host, after which the wasp larvae hatch and consume the host from the inside out. Suppose that a researcher is interested in the amount of time that a female wasp spends laying eggs on a host before moving on, a variable that often has an exponential distribution (Wajnberg and Haccou, 2008). We will put this kind of time-to-event data into a Bayesian context, as an illustration of conjugacy. Suppose that T, time-to-event, has an exponential distribution with rate parameter θ. Suppose we observe a sample of values for T, say $T_1, T_2, \ldots, T_n$. Suppose we presume that θ has a gamma prior distribution with shape parameter α' and scale parameter β'. This implies that the prior expected value of θ is

$$E'[\theta] = \frac{\alpha'}{\beta'} = \frac{0.5}{0.5} = 1.0$$

The prior variance is

$$V'[\theta] = \frac{\alpha'}{\beta'^2} = \frac{0.5}{0.5^2} = 2.0$$

The likelihood function for the sample is

$$L\left[T_1, T_2, \ldots, T_n \mid \theta\right] = (\theta)^n e^{-\theta \sum_{i=1}^{n} T_i}$$

It turns out that the posterior distribution for θ is also gamma, with shape and scale parameters:

$$\alpha'' = \alpha' + n$$

$$\beta'' = \beta' + \sum_{i=1}^{n} T_i$$

Suppose $\alpha' = 0.5$ and $\beta' = 0.5$. Also suppose after $n = 50$ observations, $\frac{1}{n} \sum_{i=1}^{n} T_i = 9.80$, and the standard deviation was 9.38. The posterior parameters are thus:

$$\alpha'' = \alpha' + 50 = 0.5 + 50 = 50.5$$

$$\beta'' = \beta' + \sum_{i=1}^{n} T_i = 0.5 + 9.80 = 10.30$$

The posterior expected value of θ is

$$E''[\theta] = \frac{\alpha''}{\beta''} = \frac{50.5}{10.30} \approx 4.903$$

And posterior variance is

$$V''[\theta] = \frac{\alpha''}{\beta'^2} = \frac{50.5}{10.30^2} \approx 0.476$$

The data shifted the central tendency of the potential values for θ, and reduced the uncertainty about the parameter's values. The prior (central) 95-percent range of possible values for θ is approximately (0.00098, 5.02389). The posterior range is (3.47609, 6.05091). The posterior range is sometimes referred to as a (central) credible interval (Winkler, 1972). A classical 95-percent confidence interval for the parameter, based solely on the data, would be (7.13, 12.47).

Bayesian Regression Analysis

A core principle of behavioral ecology is that the environment can influence behavior, but the direction and extent of this relationship is not always clear. Bronikowski and Altmann (1996) investigated whether interannual variation in the behavior of yellow baboons (*Papio cynocephalus*) could be predicted by weather patterns. They observed three groups of baboons over several years each, and for each year recorded the average distance traveled per day and the average daily time budget. Time budgets consisted of the percentage of time that adult female baboons spent feeding, moving, resting, or socializing. They also recorded several meteorological variables for each year, such as temperature and rainfall.

We will use the following example loosely based on their work to illustrate Bayesian regression. In this example, there are four continuously valued regressors that relate to environmental conditions: Average annual rainfall (mm), the coefficient of variation (CV) of annual rainfall, the minimum temperature (°C), and maximum temperature (°C). In this example, suppose that the three groups of baboons were observed over a period of 10 years. The experimental unit is the group, and the observational unit is the year. We will assume a multivariate normal prior for the regression coefficients.

Figure 10.1 shows R code for computing the posterior hyperparameters for the model.

Figure 10.2 shows some selected outputs. The vectors lowlim and upplim are the limits of the 95-percent credible intervals for the regression coefficients.

The prior hyperparameter vector, $\beta' = [1\ 1\ 1\ 1\ 1]^{\mathrm{T}}$ was chosen arbitrarily. Figure 10.3 shows a plot of the Bayesian posterior and ordinary least squares predicted values plotted against the actual observed values of response = Distance.

As in the case of any Bayesian inference, the choice of prior hyperparameters greatly affects the resulting posterior distribution. However, when used in a sequential fashion, the effect of the prior will be reduced. That is, after an initial set of data are used to evaluate posterior values, those posterior hyperparameters may become the new priors.

Then when more data become available, those new priors will get updated by computing the new posterior values.

Markov Chain Monte Carlo

As a simple example of how MCMC approximates the posterior distribution, consider a beta/binomial conjugate pair, with

$$\alpha' = 5$$

$$\beta' = 1$$

$$n = 50$$

$$y = 40$$

```
options(na.action=na.exclude)
#
# setwd tells R which folder to use for this script
#

setwd("C:\\Users\\SMFSBE\\Statistical Data & Programs")

df1 <-read.csv("20170917 Example 10.3 Amboseli Baboons.csv")
#
# Variables:
# Year
# AveRain
# CVRain
# MinTemp
# MaxTemp
# Size
# Distance
# Feeding
# Moving
# Socializing
# Resting
#
attach(df1)
lowlim <-c()
upplim <-c()
nparms <-5
response <-Distance
nupre <-1
etapre <-1
betapre <-c(1,1,1,1,1)
Inparms <-diag(nparms) # nparms x nparms Identity Matrix
Apre <-Inparms
Apreinv <-solve(Apre) #when Apre = Inparms, this step is techincally unneccessary
xmat <-as.matrix(cbind(AveRain,CVRain,MinTemp,MaxTemp))
nsamp <-nrow(xmat)#no. of rows in xmat
inter <-rep(1,nsamp) #column vector of all 1's for intercept estimation
xmataug <-cbind(inter,xmat) #adds intercept column and converts data frame into a matrix

xmatprime <-t(xmataug) #transpose function

xtx <-xmatprime %*% xmataug #matrix multiplication
Apost <-solve(Apre+xtx)
betapost <-Apost%*%(Apreinv%*%betapre + xmatprime%*%response)#Bayesian posterior

SSQpost <-t(response -xmataug%*%betapost)%*%(response-xmataug%*%betapost)
betadist <-t(betapre -betapost)%*%Apreinv%*%(betapre -betapost)
prodpost <-nupre*(etapre**2) + SSQpost + betadist
nupost <-nupre + nsamp
etapost <-sqrt(prodpost/nupost)
siginv <-drop((1/etapost**2))*(Apreinv + xtx)# you have to use the drop() funtion to let R know
1/etapost**2 is a scalar
xtxinv <-solve(xtx)
betals <-xtxinv%*%(xmatprime%*%response) #OLS estimate
#
# This loop computes individual 95% credible intervals
# for each of the posterior regression coefficients
#
for (i in 1:5) {
 bwlim[i] <-betapost[i] -qt(p=0.975,df=nupost)*(etapost/sqrt(nupost))*sqrt(Apost[i,i])
 upplim[i] <-betapost[i] + qt(p=0.975,df=nupost)*(etapost/sqrt(nupost))*sqrt(Apost[i,i])
 }
#
# note that solve can also be used to solve systems
# if a left-hand side vector is also given as an input to the function
#
#
# The posterior hyperparameters are:
# nupost
# etapost
# Apost (nparms x nparms)
# betapost (nparms x 1)
```

FIGURE 10.1
R code for Bayesian regression.

OUTPUT

```
> lowlim
[1]-0.409534729 -0.002593957 -2.017696897  0.186046849 -0.274314892
> upplim
[1] 1.73724274 0.00263902 0.02998112 1.04221276 0.07452165
> betapost [,1]
inter         6.638540e-01
AveRain       2.253136e-05
CVRain       -9.938579e-01
MinTemp       6.141298e-01
MaxTemp      -9.989662e-02
> betals[,1]
inter        -12.185768299
AveRain        0.002912948
CVRain       -17.988986970
MinTemp        1.842318701
MaxTemp        0.075704829
```

FIGURE 10.2
Selected output from Bayesian regression.

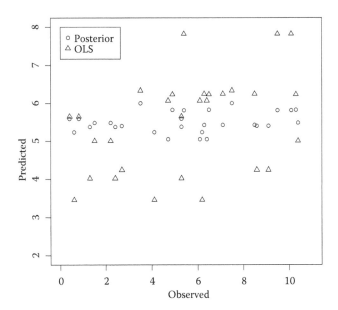

FIGURE 10.3
Plot of predicted and actual distance values.

That is, the parameter, p, has a beta prior distribution with hyperparameters α' and β'. The data are $n = 50$ trials, and $y = 40$ successes. The posterior hyperparameters are

$$\alpha'' = \alpha' + n = 5 + 50 = 55$$

$$\beta'' = \beta' + n - y = 1 + 50 - 40 = 11$$

The prior mean for p is

$$E'[p] = \frac{\alpha'}{\alpha' + \beta'} = \frac{5}{5+1} \approx 0.83$$

The posterior mean of p is

$$E''[p] = \frac{\alpha''}{\alpha'' + \beta''} = \frac{55}{55+11} \approx 0.83$$

Figure 10.4 shows the code for implementing the MCMC (Metropolis–Hastings) algorithm.

Figure 10.5 shows the histogram of the posterior samples of p.

A 95-percent credible interval for p is (0.6965, 0.8904). The arithmetic average of the posterior sample is 0.8107. Since random samples are being generated, results will vary from run to run.

Figure 10.6 illustrates yellow baboons socializing.

Theoretical Aspects

Bayesian Regression Analysis

For the usual multiple regression analysis, the model is

$$y = X\beta + \varepsilon$$

where y is a $n \times 1$ column vector of response values, X is an $n \times p$ matrix of regressor variable values, β is a $p \times 1$ column vector of unknown coefficients (or parameters) and ε is an $n \times 1$ vector of normally distributed noise values with mean 0 and some constant variance (usually). The sample size is n and the number of regressors is p.

```
options(na.action=na.exclude)
#
# setwd tells R which folder to use for this script
#
setwd("C:\\Users\\SMFSBE\\Statistical Data & Programs")

#
# This is an MCMC for a beta-binomial conjugate pair
# The conjugacy is used to illustrate how MCMC yields approximately
# the correct posterior distribution
#
pstore <- c()
alphapri <- 5
betapri <- 1
meanpri <alphapri / (alphapri + betapri)
sampsize <- 50
y <- 40
phat <- y / sampsize
p0<-0.80
nmcmc <- 2000
for (i in 1:nmcmc) {
 p1<-min(abs(rnorm(n=1,mean=p0,sd=0.05)))  #use folded & truncated normal to generate p1
 #                                 insures p1 will be in the interval (0,1)
 num <- dbeta(x=p1,shape1=alphapri,shape2=betapri)*dbinom(x=y,size=sampsize,prob=p1)
 den <- dbeta(x=p0,shape1=alphapri,shape2=betapri)*dbinom(x=y,size=sampsize,prob=p0)
 rat <- num / den
 uvar <- runif(n=1,min=0,max=1)
 if (rat >= uvar) {
  pstore[i] <- p1
  }
 else {
  pstore[i] <- p0
  }
 p0 <- pstore[i] #change p0 to current value
 }
#
#
#
alphapos <- alphapri + sampsize
betapos <- betapri + sampsize - y
meanpos <- alphapos / (alphapos + betapos)
hist(pstore)
lowlim <- quantile(x=pstore,probs=c(0.025),type=8)

medprob <- quantile(x=pstore,probs=c(0.50),type=8)
upplim <- quantile(x=pstore,probs=c(0.975),type=8)

df2 <- cbind(alphapri,betapri,alphapos,betapos,sampsize,y,lowlim,upplim,pstore)
write.csv(df2,"20170124 beta binomial posterior.csv")
```

FIGURE 10.4
MCMC code for beta/binomial example.

In the Bayesian context, the vector β is a multivariate set of random variables, and as such they have some joint prior probability distribution. Furthermore, the noise variance (which we are assuming here is the same for every observation of y_i) also is a random variable with some sort of prior distribution. There are only a few choices of priors that, together with a normal likelihood for the y_i, would provide a convenient conjugate pair.

Perhaps the simplest and maybe most common prior distribution for β is uniform:

$$f'\left(\beta|\sigma\right)=\frac{1}{\sigma}$$

Suppose that the prior distribution of $\frac{1}{\sigma^2}$ is a Chi-squared (special case of gamma) distribution with density:

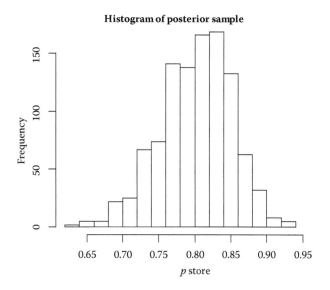

FIGURE 10.5
Histogram of posterior sample of *p*.

FIGURE 10.6
Yellow baboons (*Papio cynocephalus*).

$$f'\left(\frac{1}{\sigma^2}\right) = \frac{(1/2)^{\frac{v'}{2}}}{\Gamma\left(\frac{v'}{2}\right)}\left(\frac{1}{\sigma^2}\right)^{\frac{v'}{2}-1}\exp\left(-\frac{1}{2}\left(\frac{1}{\sigma^2}\right)\right)$$

Assuming a normal likelihood for the data vector y, then the joint posterior distribution for $(\beta, 1/\sigma^2)$ is

$$f''\left(\beta, \frac{1}{\sigma^2}\right) = \frac{1}{2\pi^{\frac{p}{2}}}\frac{2}{\Gamma\left(\frac{v''}{2}\right)}\sigma^{-p}\exp\left(-\frac{\left[v''\hat{\sigma}^2\right]+(\beta-b)^T X^T X(\beta-b)}{2\sigma^2}\right)$$

The updated (posterior) parameter v'' is $v' + n$. The vector b is the least squares multiple linear regression estimate

$$b = \left[X^T X\right]^{-1} X^T y$$

The marginal posterior density of β would be the density of the coefficient vector, "averaged" over all the possible values of σ:

$$g''_{\beta}\left(\beta|v''\right) = \int_0^\infty f''_{\beta}\left(\beta, \frac{1}{\sigma^2}\middle|v''\right)d\sigma$$

It turns out that this marginal posterior density is a multivariate Student's t (Roth, 2013):

$$g''_{\beta}\left(\beta|v''\right) = \frac{\Gamma\left(\frac{v''+p}{2}\right)}{\Gamma\left(\frac{v''}{2}\right)}\frac{1}{(v''\pi)^{\frac{p}{2}}}\frac{1}{|\hat{\Sigma}|}\left(1+\frac{1}{v''}(\beta-b)^T\hat{\Sigma}^{-1}(\beta-b)\right)^{-\left(\frac{p+v}{2}\right)}$$

The matrix $\hat{\Sigma}$ is the estimated covariance matrix for the estimates b.

A Slightly More Complicated Model

Let the prior for the coefficient vector β be a multivariate normal, with mean vector and covariance matrix:

$$E[\beta] = \beta'$$

$$Cov[\boldsymbol{\beta}] = \sigma^2 A'$$

$\boldsymbol{\beta}'$ is a $p \times 1$ vector, and A' is a $p \times p$ matrix.

Then let noise standard deviation, σ, have prior density:

$$g(\sigma) = \frac{2}{\Gamma\left(\dfrac{v'}{2}\right)} \left(\frac{v'\eta'^2}{2}\right)^{v'/2} \frac{1}{\sigma^{v'+1}} \exp\left(-\frac{v'\eta'^2}{2\sigma^2}\right)$$

This is referred to as an inverse gamma density (Judge et al., 1985). Combining these with a normal likelihood function yields posterior parameters:

$$\boldsymbol{\beta}'' = (A'^{-1} + X^T X)^{-1}(A'^{-1}\boldsymbol{\beta}' + X^T Xb)$$

But $b = [X^T X]^{-1} X^T y$, the usual least squares estimates of the regression parameters, so

$$\boldsymbol{\beta}'' = (A'^{-1} + X^T X)^{-1}(A'^{-1}\boldsymbol{\beta}' + X^T y)$$

The updated hyperparameters for σ are given by

$$v''\eta''^2 = v'\eta'^2 + (y - X\boldsymbol{\beta}'')^T(y - X\boldsymbol{\beta}'') + (\boldsymbol{\beta}' - \boldsymbol{\beta}'')^T A'^{-1}(\boldsymbol{\beta}' - \boldsymbol{\beta}'')$$

$$v'' = n + v'$$

The matrix A' is updated as

$$A'' = (A'^{-1} + X^T X)^{-1}$$

Since σ is assumed to be a random variable, the covariance matrix for the posterior distribution of β is also random.

Both β and σ are considered, in the Bayesian context, to be random variables. They have a joint prior and a joint posterior distribution. The multivariate prior normal distribution for β is conditional on having the value of σ. The joint posterior density would be

$$g''_{\beta,\sigma}\left(\boldsymbol{\beta}, \sigma | \boldsymbol{\beta}'', A'', v'', \eta''\right) = g''_{\beta}\left(\boldsymbol{\beta} | \boldsymbol{\beta}'', \sigma, A''\right) g''_{\sigma}\left(\sigma | , v'', \eta''\right)$$

It turns out that this marginal posterior distribution has the form of a multivariate Student's t:

$$g''_\beta\left(\beta|\beta'',A''\right)=\frac{\Gamma\left(\dfrac{v''+p}{2}\right)}{\Gamma\left(\dfrac{v''}{2}\right)}\frac{1}{\left(v''\pi\right)^{\frac{p}{2}}}\frac{1}{|\Sigma''|}\left(1+\frac{1}{v''}\left(\beta-\beta''\right)^T\Sigma''^{-1}\left(\beta-\beta''\right)\right)^{-\left(\dfrac{p+v}{2}\right)}$$

and

$$\Sigma''^{-1}=\frac{1}{\eta''^2}(A'^{-1}+X^TX)=\frac{1}{\eta''^2}A''^{-1}.$$

This posterior density can be used to create a p-dimensional 95-percent credible hyperellipsoid. Oh really? A hyperellipsoid? That is about as easy to comprehend as Etruscan. Rather than trying to get a region of uncertainty about all the parameters simultaneously, it is perhaps more instructive to get intervals for each parameter separately. A central 95-percent credible interval for each coefficient can also be constructed. If β_i represents the ith component of the coefficient vector (i.e., the coefficient for regressor X_i), and β''_i is the ith component of the posterior mean vector β'', then a $(1-\alpha)100\%$ credible interval for β_i is (Judge et al., 1985):

$$\beta''_i\pm t_{1-\frac{\alpha}{2}}(v'')\frac{\eta''\sqrt{c_{ii}}}{\sqrt{v''}}$$

where c_{ii} is the ith diagonal element of $(A'^{-1}+X^TX)^{-1}$, and $t_{1-\frac{\alpha}{2}}(v'')$ is the $\left(1-\dfrac{\alpha}{2}\right)100^{th}$ percentile of a (univariate) Student's t distribution with v'' degrees of freedom.

An Afterword about Bayesian Methods

Unlike classical frequentist approaches to inference, Bayesians incorporate prior information about the unknown parameters in the form of a probability distribution. The prior distribution encapsulates the best information about the parameters prior to analyzing new data. Like any model, the choice of prior can be critical in obtaining inferences about where the parameter values actually lie. Perhaps the greatest advantage to the Bayesian approach is its ability to provide updated information when additional data are obtained. That is, initially there is the prior distribution. Then some data are used to update that prior into what is called the posterior distribution. The new posterior distribution may become the new prior, which in turn may be updated if and when new data are obtained. Perhaps Bayesian methods are most appropriate when reliable prior information about parameters is available, or when data are obtained sequentially. So, perhaps several previous studies have been performed, and multiple estimates of some parameter have been obtained. The distribution of those estimates could form a prior distribution. If data are obtained in a sequential fashion, so that continual updating of parameter estimates might be performed, then Bayesian methods offer a natural means of updating.

Key Points for Chapter 10

- Both Bayesian and classical frequentist approaches to inference make use of the likelihood function, which describes the plausibility of obtaining the data under certain "hypothetical" conditions.
- Bayesian methods add another feature, called the *prior distribution*.
- The prior describes the degree of certainty about the possible ranges of unknown parameters before any new data are incorporated into analyses.
- Certain combinations of parametric forms of prior distributions and likelihoods yield posterior distributions having the same form as the prior; these combinations are called *conjugate pairs*.
- Bayesian methods can be applied to estimating parameters for predictive equations, such as multiple linear regression.
- Even when the forms of prior distributions and likelihood functions do not form conjugate pairs, the posterior distributions can often be "estimated" using simulation-based techniques known as *Markov Chain Monte Carlo (MCMC) methods*.

Exercises and Questions

1. Using the data in "20170119 Example 10.3 Amboseli Baboons.csv," perform Bayesian regression using each of the response variables. Compare the Bayesian predicted values to the OLS predictions. Explain the differences.

2. Try the same exercise as in (1), but instead of each response, use the values for the first principal component, and the first linear discriminant function (recall that there were three groups of baboons).

3. With the same dataset as in (1), instead of using the prior mean vector $\beta' = [1\ 1\ 1\ 1\ 1]^T$, try replacing it with the OLS coefficient estimates.

4. Modify the MCMC code in Figure 10.4 to perform an MCMC analysis with a Poisson/gamma conjugate pair.

11

Decision and Game Theory

General Ideas

In behavioral ecology, as in most scientific fields, empirical studies can be complemented by theoretical models that provide a mathematical description of natural phenomena. Such models necessarily make some simplifying assumptions, but within the context of these assumptions, they can be a powerful tool for generating a broader theoretical framework to explain existing empirical observations, or for generating predictions that can be tested in subsequent empirical studies. In behavioral ecology, questions of interest typically center around why animals adopt a particular behavioral strategy or tactic in a particular context. Therefore, many theoretical models in behavioral ecology rely on a branch of mathematics known as *decision theory*, which is concerned with how individuals make choices. Optimal foraging theory is a classic, well-known decision theory application that seeks to describe how individual animals make decisions about foraging for resources. Most contemporary models in behavioral ecology focus on game theory, a subset of decision theory that is concerned with interactions between individuals whose decisions affect one another. Decision theory and game theory often involve probabilistic conditions, which is why we have chosen to include them in this book on statistical methods. This chapter will be different from most of the other chapters in this book. Rather than focusing on methods for analyzing empirical data, this chapter will provide a brief introduction to theoretical models in behavioral ecology. It is not intended to equip you to create new theoretical models, but is rather intended to serve as a primer that will illustrate the basic concept of a theoretical model to the novice reader, and will hopefully make some of the simpler models in the literature a little less opaque. In this chapter, we discuss models with discrete choices and discrete states of nature, as well as models with discrete choices and continuous states of nature, using examples drawn from optimal foraging theory. We also discuss a well-known game-theoretic model, the hawk–dove model, to illustrate some of the concepts involved in modeling interactions between individuals.

In human economic decisions, the benefit is usually described in terms of monetary value. In behavioral ecology, benefit should ideally always be defined in terms of gains in fitness (lifetime reproductive success), as that is ultimately what determines the adaptive value of a behavior. However, sometimes a proxy for fitness (such as energy gain) may be used. In the context presented, the benefit function should be defined so that higher values are better than lower values.

Examples with R Code

Discrete Choices, Discrete States

Imagine that the matriarch of family of desert-dwelling elephants (*Loxodonta africana*) must choose which of three watering holes to visit. At any given hole, there are three possible "states of nature"; the hole could be dry with no water or food, the hole could have water but no food, or the hole could contain both water and food. Let's assume that all choices are mutually exclusive and exhaustive. That is, the matriarch cannot choose two holes simultaneously. Likewise, the state of any given hole can only be exactly one of the three possibilities we have outlined. Label the three holes C1, C2, and C3. Label the three possible states as follows: The hole is dry (S1), the hole has water only (S2), or the hole has water and food (S3). The benefit for choosing any given hole is a function of both the hole that is chosen (C) and the state that the chosen hole happens to be in (S). Symbolize the benefit function as $B(C_i, S_j)$, $i = 1$, 3 and $j = 1$, 3. If the benefit function is negative, then this means that there is a net cost to choosing that particular hole in that particular state. In order to select the "best" choice, there must be some strategy based on the benefit function $B(C_i, S_j)$. Strategies usually involve maximizing or minimizing something. The decision maker (in this case, the elephant matriarch) could select a strategy to minimize the maximum cost (called the minimax decision paradigm) or maximize the minimum benefit (called the maximin decision paradigm). If benefit (or cost) is strictly a function of state, then hole choice would not matter. However, suppose state is actually a random variable, and its distribution depends on the hole choice. Let the reward for S1, $B(S1) = -100$, $B(S2) = 100$, and $B(S3) = 200$. Suppose the probabilities for each of these states is given in Table 11.1.

The expected benefit would be $p_{ij}*B(C_i, S_j)$, where p_{ij} is the probability of being in state S_j at hole C_i. Consider the strategy of selecting the hole that maximizes the minimum expected net benefit. Table 11.2 shows the expected net benefit for each C_i, S_j combination, and the overall expected net benefit for each hole choice.

TABLE 11.1

State Probability Distributions: Elephants
and Watering Holes

		States		
		Dry	Water	Water & Food
		S1	S2	S3
Hole Choice	C1	0.40	0.35	0.25
	C2	0.30	0.40	0.30
	C3	0.20	0.30	0.50

TABLE 11.2

Expected Benefit: Benefit a Function of State Only

		States				
		Dry	Water	Water & Food		
		S1	S2	S3	MinEB $(S_j|C_i)$	EB (C_i)
Hole Choice	C1	−40	35	50	−40	45
	C2	−30	40	60	−30	70
	C3	−20	30	100	**−20**	**110**

Clearly, choice C3 is best, whether maximizing the minimum expected
benefit for any given state, MinEB$(S_j|C_i)$, or maximizing the expected ben-
efit over all states, EB(C_i). The function over which the maximization is performed is called the *objective* function (Reklaitis et al., 1983).

Suppose that the benefits depend on both state and choice. Perhaps the cost
for a dry hole is partly dependent on the distance to that hole, and the benefits for a "water only" hole or a "water and food" hole are partly dependent
on the quantity and quality of available water or food. Table 11.3 shows some
benefits that vary both with state and hole choice.

Assume the same probability distributions shown in Table 11.1. Then the
expected benefits are given in Table 11.4.

TABLE 11.3

Rewards that Are Dependent on State and Choice

		States		
		Dry	Water	Water & Food
		S1	S2	S3
Hole Choice	C1	−100	300	500
	C2	−200	200	400
	C3	−400	100	300

TABLE 11.4

Expected Benefits: Benefit Dependent on State and Choice

		Dry	Water	Water & Food	MinEB $(S_j|C_i)$	EB (C_i)
		S1	S2	S3		
	C1	−40	105	125	−40	190
Hole Choice	C2	−60	80	120	−60	140
	C3	−80	30	150	−80	100

Given the benefit function in Table 11.3, with the probability distributions of Table 11.1, the optimal choice in terms of either $MinEB(S_j|C_i)$ or $EB(C_i)$ is now C1.

Finally, consider the benefit function in Table 11.5, and the state probability distributions in Table 11.6. Now the benefits and probabilities have changed.

Now the cost for "Dry" at hole choice C1 is worse than at hole choice C2, but the probability of "Dry" is lower at choice C1 than at choice C2. Furthermore, the probability of "Water & Food" at hole choice C1 is now higher than it is at C2. Table 11.7 shows the expected benefits.

C1 is now the optimal choice in terms of $EB(C_i)$, but C2 is optimal in terms of $MinEB(S_j|C_i)$.

TABLE 11.5

Alternate Benefits that Are Dependent on State and Choice

		Dry	Water	Water & Food
		S1	S2	S3
	C1	−300	300	500
Hole Choice	C2	−200	200	400
	C3	−400	100	300

TABLE 11.6

Alternate Benefits that Are Dependent on State and Choice

		Dry	Water	Water & Food
		S1	S2	S3
	C1	0.25	0.35	0.40
Hole Choice	C2	0.30	0.40	0.30
	C3	0.20	0.30	0.50

TABLE 11.7

Expected Benefits: Benefit Dependent on State and Choice

		States				
		Dry	Water	Water & Food		
		S1	S2	S3	MinEB $(S_j \| C_i)$	EB (C_i)
	C1	−75	105	200	−75	**230**
Hole Choice	C2	−60	80	120	**−60**	140
	C3	−80	30	150	−80	100

In summary, there are several points:

1. In this decision paradigm, there are a discrete number of mutually exclusive choices, or actions, for the decision-maker.
2. The "states of nature" are also discrete and mutually exclusive, and they completely define all possible states.
3. The probabilities of states can vary with choice.
4. The benefit is in general a function of both choice and state.
5. There are many possible strategies for making an optimal choice.
6. The optimal choice not only depends on the state distributions and net benefit function, but on the choice of objective function (e.g., MinEB or EB).

Discrete Choices, Continuous States: Reward and Cost as a Function of Choice

It is possible for choices to be discrete, but for states to lie on a continuum. Kacelnik (1984) described choices that European starlings (*Sturnus vulgaris*) must make in terms of the number of leatherjackets (cranefly larvae, Tipulidae) to carry back to their nestlings. Davies et al. (2012) described the cost to the starlings in terms of time to search and time to travel. They also describe "profitability" in terms of benefit and cost; namely, profitability is benefit per unit cost:

$$P = \frac{B}{C}$$

So, if the "cost" is the time required to obtain and deliver the load, and the benefit is the caloric content (kcal), it seems that the optimal choice for the starling is to maximize the profitability ratio.

Suppose that *time* is a normally distributed random variable with mean (expected value) proportional to *load*:

$$\mu = E[time] = \alpha * load$$

And a constant standard deviation, σ. Assume again that the benefit for a given load is

$$B = K * load$$

where K is kcal per leatherjacket.

The profitability is

$$P = \frac{K * load}{time}$$

and

$$Pr\left\{\frac{K * load}{time} \le r\right\} = Pr\left\{time \ge \frac{K * load}{r}\right\} = 1 - Pr\left\{time \le \frac{K * load}{r}\right\}$$

So, the distribution of profitability, P, can be derived by knowing the distribution of *time* and the values of K and *load*.

As an example, suppose $\alpha = 2.0$, $\sigma = 0.3$. Figure 11.1 shows the probability density of time, and Figure 11.2 shows the cumulative distribution function of profitability for load = 4.5, $K = 10$.

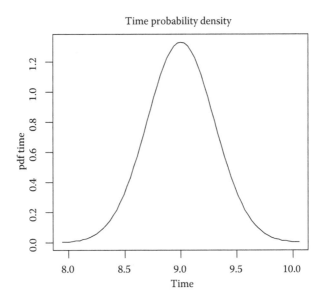

Time probability density

FIGURE 11.1
Probability density of *time*: $\alpha = 2.0$, *load* = 4.5.

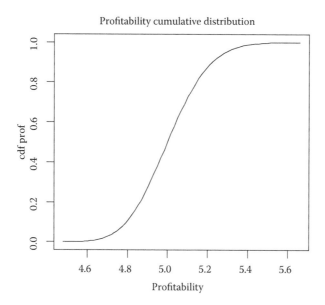

FIGURE 11.2
Cumulative distribution function of profitability.

In this case, the median profitability is 5.0 kcal per unit time, and occurs at *time* = 9.0 = α * *load* = μ (mean *time*). In fact, since the expected value of time is α * *load*, then, on the average, we might expect profitability to be

$$E[P] = \frac{K * load}{E[time]} = \frac{K * load}{\alpha * load} = \frac{K}{\alpha} = \frac{10}{2} = 5$$

This is regardless of the *load*.

It appears that expected profitability is the same for all loads. However, even if load is fixed, time is a random variable, so that the actual expected profitability is not simply $\frac{K}{\alpha}$. Figure 11.3 shows R code for computing the cdf of profitability. Table 11.8 shows the 5th and 95th percentiles of the profitability for loads ranging from 1 to 10. As load increases, the 5th percentile of profitability increases, but the 95th percentile decreases.

So, if the birds were trying to make "optimal" choices for load size, they might conclude the following:

1. To maximize the smallest profitability expected (5th percentile), choose a large load.

2. To maximize the largest profitability expected (95th percentile), choose a small load.

3. To maximize a middle value of profitability (median), choose any load.

```
setwd("C:\\Users\\SMFSBE\\Statistical Data & Programs")
#df1 <-read.csv("20170213 Example 11.2 Load and Profitability Inputs.csv")
#attach(df1)
#
# Optionally read parameters K, alpha, sigma, load from file
# or simply assign values inside script
#
K <-10
alpha <-2.0
sigma <-0.3
load <-4
mutime <-alpha*load
ltime <-max(mutime -3.5*sigma,0)
utime <-mutime + 3.5*sigma
time <-seq(from=ltime, to=utime, by=0.1*sigma)
size <-length(time)
pinc <-0.1*sigma*K*load/(utime*ltime)#this gives the same numberof prfo values as time
values
prof <-seq(from=K*load/utime,to=K*load/ltime,by=pinc)
pdftime <-dnorm(time,mean=mutime,sd=sigma)
cdftime <-pnorm(time,mean=mutime,sd=sigma)
cdfprof <-1-pnorm(q=(K*load)/prof,mean=mutime,sd=sigma)

#plot(time,pdftime,type="l",main="Time Probability Density")
#dev.new()
#plot(prof,cdfprof,type="l",main="Profitability Cumulative Distribution",xlab="Profitability")
df2 <-cbind(time,prof,cdfprof)
pctnl <-0.05
for (i in 1:size){
if (cdfprof[i] >= pctnl-0.01 & cdfprof[i] <= pctnl+0.01){
q.i <-i
q.prof <-prof[i]
q.time <-time[i]
}
}
q.i
q.prof
q.time
write.csv(df2,"20170214 Example 11.2 CDF Profitability for Fixed Load Outputs.csv")
```

FIGURE 11.3
Profitability-CDF code.

TABLE 11.8

5th and 95th Percentiles of Profitability: $K = 10$, $\alpha = 2.0$, and Associated *Time*

Load	5th Percentile Profitability	Time 5th Percentile	95th Percentile Profitability	Time 95th Percentile
1	4.0034	1.19	6.6954	1.92
2	4.4437	3.31	5.7325	4.27
3	4.6164	5.37	5.4674	6.36
4	4.7061	7.40	5.3357	8.39
5	4.7676	9.43	5.2681	10.42
6	4.8118	11.46	5.2275	12.45
7	4.8343	13.46	5.1899	14.45
8	4.8522	15.46	5.1629	16.45
9	4.8666	17.46	5.1425	18.45
10	4.8860	19.49	5.1342	20.48

Discrete Choices, Continuous States: An Inverted Problem

Here we will invert the starling problem, making *load* the random variable. Suppose that the expected time spent obtaining and delivering the leatherjackets can be described in terms of the numbers of leatherjackets to be caught (the *"load"*). Suppose further that the expected time is an exponential function:

$$time = \exp\left(\frac{load}{\gamma} - \ln\delta\right)$$

Or if time is fixed, then

$$load = \gamma * \left(ln(time) + ln(\delta)\right)$$

As before, let the caloric content (in kcal) per leatherjacket be, on the average, K. The greater the load, the higher the caloric value, but the longer the time required to obtain it. Thus, profitability is

$$P = \frac{R}{C} = \frac{K * load}{time} = \frac{K * load}{\exp\left(\dfrac{load}{\gamma} - ln\delta\right)}$$

The natural logarithm of profitability, P, is

$$lnP = lnK + ln(load) - \frac{1}{\gamma}load + ln\delta$$

Take the derivative of *lnP* with respect to load, set the derivative equal to 0 and solve for load. This gives the optimal load, which maximizes profitability P:

$$\frac{dlnP}{dload} = \frac{1}{load} - \frac{1}{\gamma} = 0$$

So, the optimal load is simply γ. In this model, the only parameter that affects optimal choice of load is γ. The time for the optimal load, continuing with the exponential model, would depend on the parameter δ:

$$time* = \exp\left(\frac{\gamma}{\gamma} - ln\delta\right) = \exp(1 - ln\delta)$$

Finally, the profitability at optimal load would be

$$P* = \frac{K * \gamma}{\exp(1 - ln\delta)}$$

Of course, both parameters δ and γ may be functions of travel distance, among other things. For simplicity, we will assume that δ is actually a "known" constant. If time was a fixed quantity, t, the parameter γ could be solved for in terms of t and *load*:

$$\gamma = \frac{load}{\ln t + \ln \delta}$$

Now assume that load is a random variable, with a Poisson distribution:

$$Pr\{load = l\} = \frac{(\theta)^b e^{-\theta}}{l!}$$

where $\theta = \gamma^*(\ln t + \ln \delta)$.

In other words, it is not known with certainty what load a starling may choose for a given foray.

The expected value of load is

$$E[load] = \theta = \gamma * (\ln t + \ln \delta)$$

The expression we gave for profitability in terms of load and γ is a very nonlinear function of γ. Since load is a random variable, so is profitability, even if time is fixed. If the value of γ is known, the distribution of profitabilities could be estimated via simulation. Figure 11.4 shows R code for executing such a simulation.

```
setwd("C:\\Users\\SMFSBE\\Statistical Data & Programs")
df1 <-read.csv("20170213 Example 11.2 Load and Profitability Inputs.csv")
# Inputs:
# K : calories per item
# A : load "shift" parameter(delta)
# B : load rate(gamma)
# time : fixed time to obtain and deliver
#
attach(df1)
Prof <-c()
load <-c()
theta <-B*(log(time) -log(A))
nsize <-2000

load <-rpois(n=nsize,lambda=theta)

for (i in 1:nsize){
if (load[i] == 0){
 Prof[i] <-0
 }
else {
 Prof[i] <-K*load[i]/exp(load[i]/B -log(A))
 }
 }
hist(Prof,freq=FALSE,main="Histogram of Profitability")
dev.new()
plot(load,Prof,main="Profitability as a Function of Load")
df2 <-cbind(load,Prof)
write.csv(df2,"20170213 Example 11.2 Loads and Profitability Outputs.csv")
```

FIGURE 11.4
Load and profitability simulation code.

Figure 11.5 shows the histogram of *load*. The expected value of *load* is

$$E[load] = \theta = \gamma * (lnt + ln\delta) = 4.5 * \big(\ln(20) + \ln(0.90)\big) \approx 13.95$$

Figure 11.6 shows the histogram of profitability, with input values:

$$K = 10$$

$$\delta = 0.90$$

$$\delta = 4.5$$

$$time = 20$$

Figure 11.7 shows a plot of profitability as a function of load.
The optimal load appears to be 4.5, although the expected value of load was 13.95 ≈ 14. While the optimal load is only dependent on the value of the parameter γ, the distribution of profitability also depends on the parameter δ.

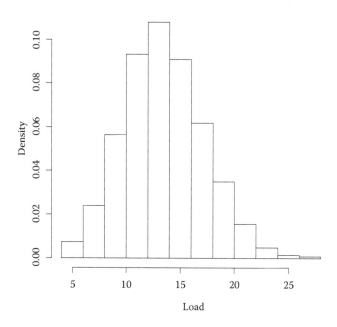

FIGURE 11.5
Histogram of load.

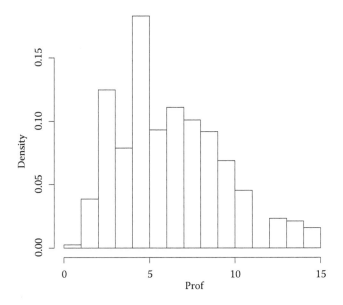

FIGURE 11.6
Histogram of profitability.

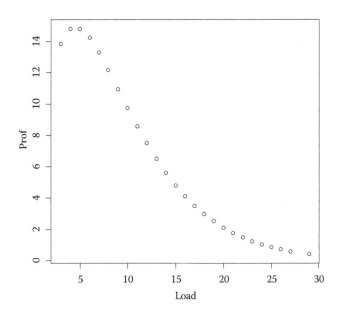

FIGURE 11.7
Profitability as a function of load.

The above sections of this chapter clearly show that optimality depends heavily on the models and assumptions made. In particular, when uncertainty about any of the variables in models exists, assumptions about their respective probability distributions can greatly affect the optimal choice.

Game Theory: Types of Games and Evolutionarily Stable Strategies

Game theory was originally developed to model human economic transactions. However, within the field of evolutionary biology, it is widely used to model the adaptive consequences of adopting a particular behavioral strategy given the strategies adopted by other individuals in the population. Each individual in the population is often referred to as a player, and the behavior adopted by a particular player is known as that player's strategy. The key difference between game theory and other types of decision–theoretic models (such as optimal foraging theory) is that in game theory, the "payoff" (net cost or benefit) for a particular strategy is dependent on the strategies adopted by other players in the population. In evolutionary game theory, the currency for payoffs is nearly always fitness, usually represented by the letter W. That is, the net benefit or cost associated with a particular strategy is defined in terms of the net effect on the player's lifetime reproductive success.

An important concept in game theory is the idea of an Evolutionarily Stable Strategy (ESS). An ESS is a strategy that once adopted by most of the individuals in the population, will provide a higher payoff on average than any rare, new strategy arising via mutation or migration. In other words, in a population of individuals playing an ESS, no new strategy can "invade" (spread within the population at the expense of the ESS), because individuals playing the ESS will have more progeny on average than individuals playing the new strategy. One caveat is that the new strategy must occur at a relatively "low" frequency at first in order for this to work; a massive influx of migrants playing a new strategy could potentially overwhelm the ESS.

For any given game, there may be one, more than one, or no possible ESSs. If an ESS does exist, it can be either "pure" or "mixed." In a pure ESS, only one strategy will exist in the population at equilibrium. However, it is still possible for more than one strategy to be a pure ESS for the same game. In this case, whichever strategy happens to become more common by random chance will rapidly go to fixation (i.e., outcompete all the other strategies until it is the only strategy present). In a mixed ESS, two or more strategies coexist at some ratio when the population is at equilibrium, and at this equilibrium ratio, the average payoff for each strategy is the same. The idea behind a mixed ESS is that a strategy's payoff is negative frequency dependent, so that when Strategy A starts to become too common, it is at a competitive disadvantage to Strategy B, and when Strategy B starts to become too common, it is at a competitive disadvantage to Strategy A. The strategies thus balance each other out and coexist at equilibrium.

Broadly speaking, there are two general classes of evolutionary game-theoretic models: Games against the field and pairwise contest games. In games against the field, there is no specific "opponent," and an individual's payoff depends on the frequencies of each strategy within the population. By contrast, in pairwise contest games, each individual in the population "plays" against a single, randomly selected opponent, and the payoff for each individual depends on what its opponent does. The pairwise contest game can be thought of as an extension of the discrete choice, discrete state problem in decision theory, where the choice is the strategy chosen by a player, and the state is the strategy chosen by its opponent. Pairwise contest games can be further broken down into symmetric and asymmetric games. In symmetric games, the expected payoffs for each strategy are the same for both players. In asymmetric games, there is an inherent difference in the roles of the two players that leads to a difference in the payoff structure that each of them experiences. That is, in an asymmetric game, the two players will not necessarily receive the same payoff even if they both adopt the same strategy.

To illustrate the basic structure of a game-theoretic model, we will focus on a well-known example of a symmetric pairwise contest game, the hawk–dove model (Maynard-Smith, 1982). In this game, players compete for a resource. A player can adopt one of two strategies: Hawk *(H)* or Dove *(D)*. The hawk strategy is to fight for the resource; the dove strategy is to not fight. If *V* (for value) represents the benefit for obtaining the resource, and *C* represents the cost associated with fighting for the resource, then the consequences, or "payoffs," for *P1* of each possible combination of strategies are summarized as follows:

$$(P1, P2) = (H, H) = \frac{1}{2}(V - C)$$

$$(P1, P2) = (D, H) = V - V = 0$$

$$(P1, P2) = (H, D) = V$$

$$(P1, P2) = (D, D) = \frac{1}{2}V$$

The rationale is that if both *P1* and *P2* choose to fight, then they each win the resource half the time and pay the price of fighting and losing half the time. If *P1* chooses strategy *D*, even though *P2* chose strategy *H*, there is no fight. Thus, *P1* gives up the resource but incurs no cost for fighting. Conversely, if *P1* chooses *H* and *P2* chooses *D*, then *P1* obtains the entire resource. Finally, if both *P1* and *P2* choose *D*, then they share the resource equally. Table 11.9 shows the strategy pairs and associated consequences for *P1* in a matrix format (known as a "payoff matrix").

An intuitive way to determine whether a particular strategy (call it "Strategy A") is an ESS is to consider what would happen if every individual in a population played Strategy A, and an individual playing any of the possible alternative strategies migrated into the population. Would the rare, new strategy have a higher payoff than Strategy A? If so, then Strategy A cannot be an ESS, because it can be successfully invaded by a rare, new strategy.

Let's apply this thought experiment to the hawk–dove game. First, we must remember that it is possible for more than one ESS to exist for the same game, so we need to examine each possible strategy in turn to determine if it is an ESS. Is hawk an ESS? Well, what would happen if a dove migrated into a population of hawks? The payoff for the dove would be $V - V = 0$, since all the other individuals it would encounter would be hawks. The hawks would mostly encounter other hawks as well, since the vast majority of the individuals in the population are still hawks. Thus, the average payoff for the hawks would be pretty close to $\frac{1}{2}(V - C)$, the payoff for a hawk encountering another hawk. If $V > C$, then $\frac{1}{2}(V - C) > 0$, and the average payoff for a hawk in a population of hawks is higher than the average payoff for a dove in a population of hawks. Thus, when $V > C$, a rare dove strategy could not invade a population of hawks, so hawk is a pure ESS.

However, if $V < C$, then $\frac{1}{2}(V - C) < 0$, and the average payoff for a hawk in a population of hawks is *lower* than the average payoff for a dove in a population of hawks. Does this mean that dove is a pure ESS when $V < C$? Not so fast! Remember that in order to be an ESS, a strategy must be able to withstand invasion *when it is played by everyone in the population*. So in order to conclude that dove is a pure ESS when $V < C$, we need to show that a population of doves cannot be invaded by a rare hawk. What would happen if a hawk migrated into a population of doves? Well, the hawk would only encounter doves (at least at first), so its payoff would be V. The doves would still mostly encounter other doves, so their average payoff would be pretty close to $\frac{1}{2}V$. Thus, dove can never be a pure ESS, even when $V < C$, because a population of all doves can always be invaded by a rare hawk.

Is there a mixed ESS when $V < C$, where there exists some ratio of hawks:doves such that the average payoff for each strategy is the same? Or is

TABLE 11.9

Hawk/Dove Game: Payoff Matrix for *P1*

		P2	
		H	**D**
P1	**H**	$\frac{1}{2}(V-C)$	V
	D	$V - V = 0$	$\frac{1}{2}V$

there no ESS at all (pure or mixed) when V < C? Imagine that V < C, and a dove migrates into a population of hawks. The dove will have a competitive advantage at first, since 0 > ½(V – C), so it will start reproducing more than the hawks (remember that "benefit" is always defined in terms of reproductive success), and doves will start increasing in frequency within the population. But wait! That means that the average payoffs for hawks and doves will start to change, since doves are no longer rare. Once the number of doves is high enough that hawks encounter doves with some regularity, then the average payoff for hawks will start to increase, since sometimes they are getting a payoff of V instead of ½(V – C). Eventually, the average payoff for a hawk will exceed the average payoff for a dove, and hawks will start to increase in frequency. Thus, when V < C, neither hawk nor dove can totally dominate the population, and there is a mixed ESS in which both strategies coexist at a particular ratio at equilibrium. Incidentally, we would have reached the same conclusion if we had started with a hawk migrating into a population of doves—the hawk strategy would spread at first, until it became too common, and doves gained a competitive advantage.

In a mixed ESS, each strategy will occur at a particular frequency at equilibrium. How can we determine the relative frequencies of each strategy at equilibrium? Let the frequency of hawks = p, and the frequency of doves = $1 - p$. The expected payoff for a hawk is thus

$$E[H] = \frac{p}{2}(V - C) + (1 - p)V$$

since the hawk will encounter other hawks with frequency p, and will encounter doves with frequency $1 - p$. Similarly, the expected payoff for a dove is

$$E[D] = p * 0 + \frac{(1 - p)}{2}V$$

At equilibrium, the average payoff for a hawk must equal the average payoff for a dove. Thus, we can set these two expressions equal to each other and solve for p, which yields

$$p = \frac{V}{C}$$

Thus, when V < C, there will be a mixed ESS with the frequency of hawks equal to $\frac{V}{C}$, and the frequency of doves equal to $1 - \frac{V}{C}$.

When there is no mixed ESS, setting the two expected payoffs equal to one another will yield a mathematically nonsensical result. For example, we already know that when V > C, there is no mixed ESS, since hawk is a pure ESS under that circumstance. If V > C, then setting $E[H] = E[D]$ and solving for $p = \dfrac{V}{C}$ implies that the proportion of hawks (p) in the population is greater than one, which is impossible. This confirms our conclusion that a mixed ESS can only exist in the hawk–dove game when V < C.

This treatment of game theoretic models in behavioral ecology is only a brief introduction to a vast field of study. Interested readers should consider reading Dugatkin and Reeve (1998) for an in-depth treatment of evolutionary game theory. For a more general coverage of game theory, see Tadelis (2013).

Theoretical Aspects

Verifying Models: Frequentist and Bayesian Approaches

Optimality of choice depends heavily on the models for benefit, cost, state, and even choices themselves. The frequentist approach relies on fitting models to data, and then deciding if the models fit well. In the elephant/water hole example, it might not be too difficult to estimate state frequencies (S1 = Dry, S2 = Water, S3 = Water & Food) for each hole (C1, C2, C3). Similarly, the frequencies with which the elephant matriarch chooses C1, C2, and C3 might also be observable. While the true benefit or cost is the impact on fitness, we will think of the cost as the distance traveled to each hole. In this simple model, the benefit will be represented implicitly as a sort of negative cost. Nevertheless, using the empirically obtained state frequencies for each hole, the hole that minimizes the maximum expected cost could be determined. Then the frequency that the elephants visit the "optimal" hole could be compared to the frequency in which they choose the "non-optimal" holes. If the optimal hole frequency is significantly greater, it would appear that the elephant matriarch is using a minimax decision paradigm.

Alternatively, a Bayesian formulation would first specify a prior distribution for states for each hole. One possibility is to have the same prior for each hole choice:

$$p_i = Pr\left\{S_i \middle| C_j\right\}, C_j, j = 1, 3$$

If the data for state frequency were of the form x_{ij} out of n_j for state S_i and hole choice C_j, and the empirical proportion of state S_i at hole C_j is

$$\hat{p}_{ij} = \frac{x_{ij}}{n_j}$$

then the likelihood function would be binomial:

$$L\left(x_{ij}\,\middle|\,n_j, S_i, C_j\right) = \binom{n_j}{x_{ij}} p_i^{x_{ij}} (1-p_i)^{n_j - x_{ij}}$$

From Bayes' theorem, the posterior probability for S_i at hole C_j is

$$Pr\{Si \mid x_{ij}, n_j, C_i\} = \frac{Pr\{S_i \mid C_j\} L(x_{ij} \mid n_j, S_i, C_i)}{\sum_{k=1}^{3} Pr\{S_k \mid C_j\} L(x_{ij} \mid n_j, S_k, C_j)}$$

$$= \frac{(p_i)\binom{n_j}{x_{ij}}(p_i)^{x_{ij}}(1-p_i)^{n_j - x_{ij}}}{\sum_{k=1}^{3} (p_k)\binom{n_j}{x_{ij}}(p_k)^{x_{ij}}(1-p_k)^{n_j - x_{ij}}}$$

By way of example, suppose the state probabilities are actually those shown in Table 11.6, and the associated benefits are given in Table 11.5. Under those conditions, the optimal choice, averaged over all states, was C1. Table 11.10 shows the numbers of days each state occurred at each hole, over a 300-day period.

The frequentist approach might be to first test the hypothesis that the table of state probabilities was actually believable, given the data. The second thing might be to estimate the expected benefit for each hole choice. The third thing might be to compute a confidence interval for expected benefit for each hole choice. Table 11.11 shows the estimated expected benefits (recall that for each hole, $n = 300$ days over which the states were observed).

From this table, it appears that the optimal hole choice is C2.

The Bayesian would take a different approach, using the data to update the prior state frequency table. The entries in the table are "prior" state probabilities, given hole, and the likelihood function for each observed count of state

TABLE 11.10

Observed Frequencies of States for Each Hole Choice

		States		
		Dry	Water	Water & Food
		S1	S2	S3
	C1	139	57	104
Hole Choice	C2	41	181	103
	C3	70	77	130

TABLE 11.11

Estimated Expected Benefits based on State Frequencies

Hole	Est. EB
C1	91.33
C2	230.67
C3	62.33

TABLE 11.12

Posterior Expected Benefits

Hole	Est. EB
C1	500.00
C2	400.00
C3	−75.38

is binomial, then the updated posterior probabilities can be used to obtain a posterior expected benefit for each hole. These are shown in Table 11.12.

Thus, the Bayesian would choose hole C1 as optimal.

Finally, the question is, "What does the matriarch actually choose most frequently?" Suppose she chooses C1 most frequently. Would that mean the Bayesian methodology was better than frequentist? Suppose she chose C2 most frequently. How would that reflect on the frequentist methodology? Finally, they could have any number of possible frequency distributions for hole choice. They could have a more or less uniform frequency of hole choice, have chosen C3 most frequently, or two holes could be tied for highest frequency. Any of these possible outcomes might lead the researcher to reevaluate the benefit function. Possibly there is another factor in determining benefit, such as the likelihood that predators (humans or lions) would be present at each of the holes.

Key Points for Chapter 11

- Animals make choices, and the choices have consequences that depend on the states of nature, or conditions that exist when the choice is made.
- The benefits and costs are quantifications of those consequences.
- Animals may or may not choose an "optimal" decision.
- Optimality of a choice can be determined by comparing the net reward possible over all possible states of nature; one paradigm

for optimality is to make the choice that maximizes the minimum reward possible.

- A more probabilistic version of max–min is to maximize the minimum expected reward, based on probability distributions for states.
- States can be discrete or continuously valued.
- The optimum choice depends heavily on the model describing benefit, cost, or the benefit per unit cost, called *profitability*.
- Game theory is related to decision theory, where the states are actually choices, called *strategies*, made by competing entities.
- A strategy that once fixed in a population cannot be invaded by a new strategy is called an Evolutionarily Stable Strategy (ESS).

Exercises and Questions

1. Compute posterior probabilities for the states of the three holes, using the frequencies in Table 11.10, and prior probabilities in Table 11.6.

2. How would you test the hypothesis that state probabilities in Table 11.6 are believable?

12

Modern Prediction Methods and Machine Learning Models

General Ideas

Do Machines Learn?

Machine learning methods are, loosely speaking, a collection of model-building methods where an algorithm determines the model used for prediction. An algorithm is a procedure or process, usually iterative in nature, that has some a priori rule indicating when the iteration is complete. The iteration must occur in such a way as to update the inputs, which in turn changes the value of some measure of desirability, called the objective function. Usually an algorithm requires some inputs, such as starting conditions. Furthermore, the rule is specified in such a way as to guarantee the process will complete. Sometimes the completion of the algorithm is called *convergence*. So, for example, consider the process of making toast. The bread is sliced and then placed into the toaster. The heating element is then activated, and will remain activated until the preset time has expired or there is manual intervention, whichever comes first. One might refer to the ending of the process when a piece of bread is successfully toasted as *convergence*. The toasting process is not exactly iterative, inasmuch as time is a continuous variable. Nevertheless, it is an algorithm. If there were no timer or intervention, then the toaster would continue to cook the toast, ostensibly forever, although most likely the toaster would break down or burst into flames, thus stopping the process. If one checked at regular time intervals to determine if the bread was sufficiently toasted, the process would be iterative. To be an algorithm, there must be absolute certainty that the process will end.

On occasion, software that implements an algorithm doesn't work right, often giving the user a message such as "algorithm failed to converge." The programmers have put into their code a failsafe mechanism that would not allow any process to continue infinitely. So, on occasion, the numerical precision of the computer is such that it cannot proceed to get closer to achieving the stopping rule. Thus, the program will artificially end the process, and tell the user that something did not go according to plan. Since all digital computers represent continuous, "floating point" numbers by a finite number of bits (a bit is a numeric value that can only be translated as either zero or one), there is actually a smallest number that the computer can represent, and a largest number. When programs give the user a message such as "algorithm failed to converge," it is often due to some computation that comes very close to the largest number the computer can represent, or due to division by a number smaller than the smallest number the computer can represent (effectively, division by zero). Sometimes the failure to converge is due to the starting or initial conditions, or some other inputs that are outside the range of inputs anticipated by the programmer.

As mentioned earlier, an algorithm is based on something called the *objective function*. This is some quantitative formula that is dependent on inputs, and whose value gets altered through the course of the algorithmic process. The stopping rule usually involves checking the objective function's value periodically, and stopping when the objective has not changed "substantially" since the last time it was checked. The inputs to the objective function are altered at each stage of the process, hopefully making the objective function's value change to some "optimal" point.

Technically speaking, least squares multiple regression is an algorithm, but the model is chosen by the analyst, and not by a computer program. Machine learning algorithms are sometimes referred to as unsupervised learning, in that the decision about model details is determined by the program, and not by the human analyst. Even Bayesian sequential analysis, while employing an algorithm, still requires human specification of prior distributions and likelihood functions. We will explore three types of machine learning algorithms. Even these are not entirely free of human intervention, but they do rely more heavily than some older methods on algorithmic determination of predictive models. The four methods discussed here are stepwise regression, (artificial) neural networks, CART, and Bayesian model averaging.

Examples with R Code

For this chapter, we will employ a simulated dataset based on a hypothetical study of elephant vocalizations. The information used to construct our example dataset was drawn from the works of Poole et al. (1988), King et al. (2010), Poole (2011), De Silva et al. (2011), and Soltis et al. (2014). While the example may not reflect more realistic questions that could be answered using the methods of this chapter, it provides a relatively simple application.

African elephants (*Loxodonta africana*) produce powerful, low-frequency vocalizations called rumbles, which are partially (although often not entirely) below the range of human hearing. Rumbles are given in a wide variety of behavioral contexts, suggesting that there may be a variety of different types of rumbles with different meanings. However, for some of these putatively different rumbles, it has yet to be determined whether the rumbles produced in different behavioral contexts are actually acoustically distinct from one another. To determine whether each putative type of rumble is in fact a distinct call, or whether the same "multi-purpose" rumble is used in several different contexts, it is necessary to test whether the acoustic features of a rumble can be used to predict the behavioral context in which it was given. Suppose an intrepid research team recorded 350 rumbles from several adult female elephants in the following behavioral contexts:

1. Let's go: Rumbles given right before the group begins to move.
2. Post-copulatory: Rumbles given by an estrus female after copulation.
3. Greeting: Rumbles given by social affiliates who are approaching each other after at least several hours apart.
4. Contact call: Rumble given by an elephant separated from its group.
5. Contact answer: Rumble given in response to a contact call, or in response to another contact answer.
6. Bee alarm: Rumbles given in response to a disturbed beehive.
7. Predator alarm: Rumbles given in response to predator such as humans or lions.

The researchers then measured the following acoustic features on each of the rumbles. The fundamental frequency is the lowest frequency at a given point in the call, and is determined by the rate at which the elephant's larynx vibrates during vocalization. The fundamental frequency may change throughout the duration of the call, producing a wavy contour line in a spectrogram. Harmonics are higher frequencies overlaid on top of the fundamental frequency. Formants are bands of harmonics that are increased in amplitude relative to other harmonics because of the resonance properties of the elephant's vocal track. If the elephant manipulates the shape of its vocal tract with its tongue or other muscles, this will change the frequencies at which the formants occur.

1. Duration: Number of seconds from the start to the end of the call.
2. F0 Start: Fundamental frequency at the beginning of the call.
3. F0 End: Fundamental frequency at the end of the call.
4. F0 Max: Highest point of the fundamental frequency contour.
5. F0 Min: Lowest point on the fundamental frequency contour.
6. Formant 1: Frequency of the lowest formant in the call.
7. Formant 2: Frequency of the second lowest formant in the call.

8. Harmonic-to-noise ratio (HNR): A measure of how periodic (i.e., tonal) the call is. It is calculated as $10*\log_{10}(H/N)$, where H=% of the total energy of the call that is found in the harmonics, and N=% of the total energy of the call that is found in between the harmonics. Higher values indicate higher tonality.

Stepwise Regression

Stepwise regression is perhaps the easiest method to explain in this chapter. The idea behind stepwise regression is that there are many potential regressors that can be used to make predictions of a single response variable, but it is not clear what combination of these regressors provides the best predictions. There are two basic flavors of stepwise regression: forward and backward. In forward stepwise regression, a single regressor is added, and then another, and another, until all the possible regressors have been tested. At each "step," an assessment is made as to whether to keep the new regressor, or exclude it, and move on to the next. Backward regression is the opposite; initially all regressors are included, and then at each step, some subset of regressors is kept and some are removed. Since there are a finite number of regressors, the stepwise procedure is guaranteed to stop, making it an algorithm.

There are many different criteria that can be used to decide whether a regressor should be entered or removed from a model. A common criterion is called Akaike Information Criterion (*AIC*) (Akaike, 1974). In general, lower values of *AIC* indicate a better model. *AIC* is computed as

$$AIC = -2\ln(L) + k * edf$$

L is the value of the likelihood function for the model. The value of edf is the "equivalent degrees of freedom," which is the degrees of freedom for the model. The value of k is a somewhat arbitrary penalty weight. If k = ln(n), then *AIC* is called *BIC* (Bayesian Information Criterion). As edf and k get larger, so does AIC, which is not desirable.

Under the assumption of a linear model, with normally distributed residuals, the *AIC* is computed as

$$AIC = \frac{RSS}{\hat{\sigma}^2} - n$$

RSS is the residual sums of squares for the model at the given step, *n* is the total sample size, and $\hat{\sigma}$ is the root mean square error from the "full" model. This is closely related to another measure of model goodness, called Mallow's *Cp* (Mallows, 1973):

$$C_p = \frac{RSS}{\hat{\sigma}^2} - n + 2p$$

The value of *p* is the number of coefficients in the current model, including the intercept term. While lower values are "better" for C_p, the ideal value is actually *p*, the number of parameters. The value of *p* is in a sense optimal for C_p, so that a model with C_p closest to *p* would be considered "best."

The stepwise approach would compute *AIC* at each step, and the algorithm would attempt to minimize *AIC*.

In this example, for the convenience of having a continuous response, HNR will be used. Essentially, the researchers are asking the question of whether the harmonics-to-noise ratio of an elephant rumble can be predicted by its other acoustic properties. The linear model to be fit can be expressed as

$$\text{HNR} = \beta_0 + \beta_1 \text{Duration} + \beta_2 \text{F0.Start} + \beta_3 \text{F0.End} + \beta_4 \text{F0.Min}$$
$$+ \beta_5 \text{F0.Max} + \beta_6 \text{Format.1} + \beta_7 \text{Format.2} + \text{noise}$$

The *step()* function in R uses an *lm()* object to perform stepwise regression. The code in Figure 12.1 shows how it is implemented. The output is given in Figure 12.2.

The final model is given in Figure 12.3. The "Estimate" column shows the estimated coefficients (including the intercept term).

One way to illustrate the quality of this model is to fit a line to the predicted vs. actual values of the response, HNR. Figure 12.4 shows the points (HNR, predicted HNR) with the regression line, and associated regression estimates and statistics.

Figure 12.5 shows the quality of the fitted versus observed HNR values both for the reduced (final) model from the stepwise procedure and the "full" model. Note how close both models are to each other.

It appears that the reduced model, having only four regressors plus an intercept, is just as good at predicting HNR as the "full" model, which has seven regressors plus an intercept. By the principle of parsimony in model-building (remember William Ockham?), the reduced model is in some sense better than the full model.

A somewhat related methodology to stepwise regression is called *all possible models regression* (Draper and Smith, 1998). The R function called dredge(), in package MuMIn implements this methodology. The idea is to fit all possible models with all combinations of regressors (and their two-way interactions),

```
setwd("C:\\Users\\SMFSBE\\Statistical Data & Programs")
df1 <-read.csv("20170221 Example 12.1 Elephant Calls.csv")
#
#
# VARIABLES:
# Call.ID
# CallType
# Duration
# F0.Start
# F0.End
# F0.Min
# F0.Max
# Formant.1
# Formant.2
# HNR
#
attach(df1)

linreg <-lm(HNR ~ Duration + F0.Start + F0.End + F0.Min + F0.Max + Formant.1 +
Formant.2)
#
# Make some plots:
#
# dev.new() allows more plots to be made without overwriting previous plots
#

summary(linreg)

steplin <-step(linreg)
regfitted <-lm(steplin$fitted.values ~ HNR)
summary(steplin)
hist(HNR)
dev.new()
plot(steplin$fitted.values,steplin$residuals)
dev.new()
plot(HNR,steplin$fitted.values,pch=1,col="black",ylab=c("fitted values"))
points(x=HNR,y=linreg$fitted.values,pch=2,col="red")
abline(a=regfitted$coefficients[1],b=regfitted$coefficients[2],lty=1,col="black")
abline(a=bigmodel$coefficients[1],b=bigmodel$coefficients[2],lty=3,col="red")
legend(x=10,y=25,legend=c("reduced v. HNR (points)","full v. HNR (points)","reduced
model","full model"),pch=c(1,2,1,2),lty=c(0,0,1,3),col=c("black","red","black","red"))
summary(regfitted)
summary(bigmodel) plot(steplin$fitted.values,steplin$residuals)
dev.new()
plot(HNR,steplin$fitted.values,pch=1,col="black",ylab=c("fitted values"))
points(x=HNR,y=linreg$fitted.values,pch=2,col="red")
abline(a=regfitted$coefficients[1],b=regfitted$coefficients[2],lty=1,col="black")
abline(a=bigmodel$coefficients[1],b=bigmodel$coefficients[2],lty=3,col="red")
legend(x=10,y=25,legend=c("reduced v. HNR (points)","full v. HNR (points)","reduced
model","full model"),pch=c(1,2,1,2),lty=c(0,0,1,3),col=c("black","red","black","red"))
summary(regfitted)
summary(bigmodel)
```

FIGURE 12.1
Stepwise regression in R.

and rank them by some criterion. The default criterion is AIC, so that dredge() would choose the model with the smallest AIC value (not the AIC coming closest to p, the number of parameters in the model). We will not present any examples using dredge(). Such methods can be helpful when the numbers of regressors is very large and there is no scientific (physical, biological, social, or psychological) reason to include or exclude any of them. However, while dredge() might provide a predictive model, it can easily overlook important relationships between responses and regressors, and is not very likely to provide insight into how the regressors may affect the response. Stepwise regression attempts to build a model one term at a time, so that each candidate model can be examined by the experimenter. The all possible models approach may miss models that are perhaps suboptimal with respect to the criterion, but may provide the researcher with some valuable insights, or at least questions.

```
> summary(linreg)

Call:
lm(formula = HNR ~ Duration + F0.Start + F0.End + F0.Min + F0.Max +
   Formant.1 + Formant.2)

Residuals:
   Min       1Q      Median       3Q        Max
-0.54778  -0.15837   0.00775    0.15150    0.57038

Coefficients:
              Estimate    Std. Error    t value    Pr(>|t|)
(Intercept)   0.0963984   0.1105833      0.872      0.384
Duration      0.0028764   0.0041429      0.694      0.488
F0.Start      0.0013377   0.0107345      0.125      0.901
F0.End        0.0062461   0.0080897      0.772      0.441
F0.Min        0.2315330   0.0098004     23.625     <2e-16 ***
F0.Max        0.2526830   0.0034173     73.942     <2e-16 ***
Formant.1     0.0998072   0.0005290    188.685     <2e-16 ***
Formant.2    -0.0100037   0.0002831    -35.337     <2e-16 ***
---
Signif. codes:  0 '***' 0.001 '**' 0.01 '*' 0.05 '.' 0.1 ' ' 1

Residual standard error: 0.2107 on 342 degrees of freedom
Multiple R-squared:  0.9983,  Adjusted R-squared: 0.9983
F-statistic: 2.883e+04 on 7 and 342 DF,  p-value: < 2.2e-16

>
> steplin <-step(linreg)
Start:  AIC=-1082.28
HNR ~ Duration + F0.Start + F0.End + F0.Min + F0.Max + Formant.1 +
   Formant.2

              Df   Sum of Sq      RSS       AIC
-F0.Start      1      0.00       15.18   -1084.26
-Duration      1      0.02       15.20   -1083.79
-F0.End        1      0.03       15.21   -1083.67
<none>                           15.18   -1082.28
-F0.Min        1     24.77       39.95    -745.57
-Formant.2     1     55.42       70.60    -546.29
-F0.Max        1    242.68      257.86     -92.93
-Formant.1     1   1580.25     1595.43     544.94

Step:  AIC=-1084.26
HNR ~ Duration + F0.End + F0.Min + F0.Max + Formant.1 + Formant.2

              Df   Sum of Sq      RSS       AIC
-Duration      1      0.02       15.20   -1085.78
-F0.End        1      0.04       15.22   -1085.38
<none>                           15.18   -1084.26
-F0.Min        1     43.84       59.02    -611.04
-Formant.2     1     59.26       74.44    -529.76
-F0.Max        1    261.32      276.50     -70.51
-Formant.1     1   1580.26     1595.44     542.94

Step:  AIC=-1085.78
HNR ~ F0.End + F0.Min + F0.Max + Formant.1 + Formant.2

              Df   Sum of Sq      RSS       AIC
-F0.End        1      0.02       15.23   -1087.22
<none>                           15.20   -1085.78
-F0.Min        1     45.52       60.72    -603.09
-Formant.2     1     61.86       77.06    -519.68
-F0.Max        1    289.07      304.27     -39.00
-Formant.1     1   1593.94     1609.15     543.93

Step:  AIC=-1087.22
HNR ~ F0.Min + F0.Max + Formant.1 + Formant.2

              Df   Sum of Sq      RSS       AIC
<none>                           15.23   -1087.22
-Formant.2     1     64.49       79.71    -509.83
-F0.Min        1     68.81       84.04    -491.34
-F0.Max        1    301.09      316.32     -27.42
-Formant.1     1   1616.66     1631.88     546.84
```

FIGURE 12.2
Stepwise regression output.

```
> summary(steplin)


Call:
lm(formula = HNR ~ F0.Min + F0.Max + Formant.1 + Formant.2)


Residuals:
    Min       1Q     Median      3Q       Max
-0.55469  -0.15202   0.01267   0.14517   0.55348


Coefficients:
                Estimate    Std.Error    t value    Pr(>|t|)
(Intercept)    0.1631938    0.0872413      1.871      0.0622
F0.Min         0.2362832    0.0059841     39.485     <2e-16***
F0.Max         0.2539246    0.0030743     82.596     <2e-16***
Format.1       0.0998174    0.0005215    191.391     <2e-16***
Format.2      -0.0100846    0.0002638    -38.225     <2e-16***
---
Signif.codes: 0'***' 0.001'**' 0.01'*' 0.05'.'    0.1''1


Residual standard error: 0.2101 on 345 degrees of freedom
Multiple R-squared: 0.9983, Adjusted R-squared: 0.9983
F-statistic: 5.74e+04 on 4 and 345 DF, p-value: < 2.2e-16
```

FIGURE 12.3
Final model from stepwise procedure.

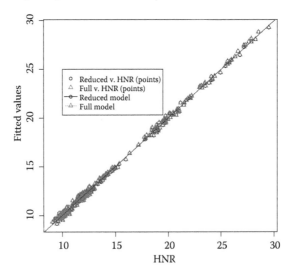

FIGURE 12.4
Fitted, or predicted, HNR values versus observed HNR.

```
> summary(regfitted)

Call:
lm(formula = steplin$fitted.values ~ HNR)

Residuals:
     Min       1Q    Median       3Q      Max
-0.55975  -0.14886  -0.01006  0.15050  0.54943

Coefficients:
               Estimate    Std. Error    t value    Pr(>|t|)
(Intercept)    0.024432      0.033672      0.726       0.469
HNR            0.998303      0.002206    452.454      <2e-16 ***
---
Signif. codes:  0 '***' 0.001 '**' 0.01 '*' 0.05 '.' 0.1 ' ' 1

Residual standard error: 0.209 on 348 degrees of freedom
Multiple R-squared:  0.9983,    Adjusted R-squared:  0.9983
F-statistic: 2.047e+05 on 1 and 348 DF,  p-value: < 2.2e-16


> summary(bigmodel)

Call:
lm(formula = linreg$fitted.values ~ HNR)

Residuals:
     Min       1Q   Median       3Q      Max
-0.57663  -0.14988  -0.00954  0.15831  0.54630

Coefficients:
               Estimate    Std. Error    t value    Pr(>|t|)
(Intercept)    0.024358      0.033621      0.724       0.469
HNR            0.998308      0.002203    453.143      <2e-16 ***
---
Signif. codes:  0 '***' 0.001 '**' 0.01 '*' 0.05 '.' 0.1 ' ' 1

Residual standard error: 0.2087 on 348 degrees of freedom
Multiple R-squared:  0.9983,    Adjusted R-squared:  0.9983
F-statistic: 2.053e+05 on 1 and 348 DF,  p-value: < 2.2e-16
```

FIGURE 12.5
Fitted versus observed HNR: Regression output.

While there are those who find the all possible model approach to have merit, we will not present anything more on this methodology.

Artificial Neural Networks

The original idea behind the artificial neural network was that it should mimic the way in which biological neurons transmit data so that the organism "learns." As it turns out, artificial neural networks do not actually mimic biological neurons in any real way. However, they are a useful tool for building a predictive model when the relationships between response and predictors are not understood. The way they work is by making guesses at how to relate a particular set of predictor or regressor variables with responses using a "training set" of data. Then, a test set can be used to see just how good the guess was. The tool is called a neural network, because there are layers

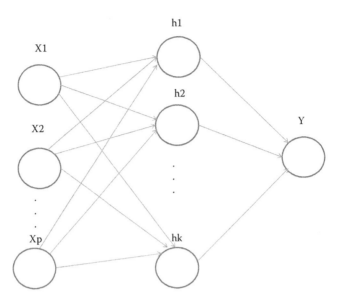

FIGURE 12.6
A conceptual neural network.

through which the information moves. The outmost layers are the inputs and the responses. Between those layers are the nerve cells. Each cell will get "stimulated" by each input, and then produce some sort of "output," that can then feed into the next layer, and so on, until the final neural layer produces a "response." The internal layers of nerve cells are called "hidden" layers. Most applications seem to do best with a single hidden layer. Each nerve cell (neuron) consists of a little function with some parameter that is estimated using the training data. In other words, the algorithm finds the "best" values of the parameters so that the predicted responses most closely match the observed responses. Figure 12.6 illustrates the concept of the artificial neural network (referred to from here on as neural networks).

Neural networks can be used in two ways. First, they can be used to classify unknown individuals into one of several a priori defined classes, as in the case of linear discriminant analyses. They can also be used as multiple regression models, to compute a predictive equation for continuously valued responses.

The little functions within each neuron are often the logistic "S" curve function (usually most appropriate when the neural net is used for classification):

$$h(x) = \frac{1}{1 + e^{-\beta^T x}}$$

However, the functions in the hidden layer(s) can be simply linear:

$$h(x) = \beta^T x$$

There is no actual restriction on the forms of the $h(x)$ functions. However, the S-shaped curves and linear curves seem to be the most common choices. The variable x is a vector of inputs from the previous layer, and β a vector of unknown "weights." The idea is that the input and response data are used to estimate the weights to minimize the error in associating the known inputs with their corresponding response values. In fact, a numerical least squares approach is often used to estimate the weights.

As an example, consider the elephant call data. Suppose we had some acoustic feature data for some calls, and we wanted to guess at what type of calls they were. Recall that there were seven types:

1. Let's go
2. Post-copulatory
3. Greeting
4. Contact call
5. Contact answer
6. Bee alarm call
7. Predator alarm call

The basic neural network R function is called *nnet()*, in package nnet. Its primary input is a formula similar to those used in the *lm()* function. If the left-hand side variable is a factor, then nnet will perform as a classifying algorithm. For each individual, *nnet()* will compute a probability that the individual belongs to each group defined by the discrete levels of the response. We have provided some code in Figure 12.7. We added a variable called "testind" as a means of identifying a training set and a test set. The training set will be used to "fit" the neural network (i.e., estimate parameter values), and the test data will be used to compare predicted versus actual call types. The parameter size is the number of nerve cells you want in the hidden layer. There is no magic formula for determining the best number. It can be empirically estimated by trying different numbers and observing how many classification errors the neural network makes with the test data. A logical input, called *skip*, tells the network to either have, in addition to connections between the input layer and hidden layer, connections between the inputs and output layer directly. The "linout" parameter indicates whether the outputs from each nerve cell should be a linear function (linout = TRUE) or a logistic function (linout = FALSE). Table 12.1 shows a table of results for the test dataset, namely the numbers of results actually associated with a particular call type versus the predicted call types. The only error made by the neural network with the test data was a contact call that was misidentified as a bee alarm call. In the code, there is a line that is "commented out," which explains how to use the neural network as a regression model. The analyses are left as an exercise for the reader.

```
setwd("C:\\Users\\SMFSBE\\Statistical Data & Programs")
df1 <-read.csv("20170301 Example 12.1 Elephant Calls.csv")
#
#
#  need to execute library(nnet) once during R session
#
#VARIABLES:
# Call.ID
# CallType
# Duration
# F0.Start
# F0.End
# F0.Min
# F0.Max
# Formant.1
# Formant.2
# HNR
# testind
#
nbin <-nrow(df1)
#testind <-rbinom(n=nbin,size=1,prob=0.1) #currently testind is an input column
attach(df1)
fCall.ID <-factor(Call.ID)
levels(fCall.ID) <-c("let's.go","post.cop","greet","cont.call","cont.answ","bee","predator")
df2 <-data.frame(cbind(df1,fCall.ID))
detach(df1)
attach(df2)
dftest <-subset(df2,testind==1)
dftrain <-subset(df2,testind==0)
fCall.ID <-factor(Call.ID)
#reg.df <-
data.frame(cbind(Duration,F0.Start,F0.End,F0.Min,F0.Max,Formant.1,Formant.2,HNR))
#
detach(df2)
attach(dftrain)
linreg <-lm(HNR ~ Duration + F0.Start + F0.End + F0.Min + F0.Max + Formant.1 +
Formant.2,data=dftrain)
# Make some plots:
#
# dev.new() allows more plots to be made without overwriting previous plots
#

summary(linreg)

    HNRnet <-nnet(fCall.ID ~ Duration + F0.Start + F0.End + F0.Min + F0.Max + Formant.1 +
    Formant.2,data=dftrain,size=3,skip=TRUE,linout=FALSE)
    #HNRnet <-nnet(HNR ~ Duration + F0.Start + F0.End + F0.Min + F0.Max + Formant.1 +
    Formant.2,data=dftrain,size=1,skip=TRUE,linout=TRUE)
    fClass.obs <-dftest$fCall.ID
    fClass.pred <-predict(HNRnet,dftest,type="class")

    table(actual=fClass.obs,predicted=fClass.pred)
```

FIGURE 12.7
Neural network code: Elephant call data.

Classification and Regression Trees (CART)

CART is a way of generating predictions using linear functions. It resembles a tree since it has stages (like stepwise regression) and at each stage the data are split into two branches. The split is based on a regressor variable that produces two models with all the remaining regressors in such a way as to minimize the root mean square error (RMSE). The splitting continues until the sample size left is too small for splitting. CART is particularly useful when there are many regressors, so that an ordinary multiple linear regression model would be highly over-parameterized. CART creates a tree, where each branch will lead to a predicted value. The R function *rpart()* in the package rpart will create this regression tree. Figure 12.8 shows code for fitting a regression tree with the elephant call data.

TABLE 12.1

Classification of Calls by the Neural Network
> table(actual=fClass.obs,predicted=fClass.pred)

	predicted						
actual	bee	cont.answ	cont.call	greet	let's.go	post.cop	predator
let's.go	0	0	0	0	2	0	0
post.cop	0	0	0	0	0	6	0
greet	0	0	0	3	0	0	0
cont.call	1	0	4	0	0	0	0
cont.answ	0	3	0	0	0	0	0
bee	5	0	0	0	0	0	0
predator	0	0	0	0	0	0	3

```
setwd("C:\\SMFSBE\\Statistical Data & Programs")
df1 <-read.csv("20170301 Example 12.1 Elephant Calls.csv")
#
#
#  need to execute library(nnet) once during R session
#
# VARIABLES:
# Call.ID
# CallType
# Duration
# F0.Start
# F0.End
# F0.Min
# F0.Max
# Formant.1
# Formant.2
# HNR
# testind
#
nbin <-nrow(df1)
#testind <-rbinom(n=nbin,size=1,prob=0.1) #currentlytestind is an input column
attach(df1)
fCall.ID <-factor(Call.ID)
levels(fCall.ID) <-c("let's.go","post.cop","greet","cont.call","cont.answ","bee","predator")
df2 <-data.frame(cbind(df1,fCall.ID))
detach(df1)
attach(df2)
dftest <-subset(df2,testind==1)
dftrain <-subset(df2,testind==0)
fCall.ID <-factor(Call.ID)
#reg.df <-
data.frame(cbind(Duration,F0.Start,F0.End,F0.Min,F0.Max,Formant.1,Formant.2,HNR))
#
detach(df2)
attach(dftrain)
linreg <-lm(HNR ~ Duration + F0.Start + F0.End + F0.Min + F0.Max + Formant.1 +
Formant.2,data=dftrain)
# Make some plots:
#
# dev.new() allows more plots to be made without overwriting previous plots
#

summary(linreg)


HNRcart <-rpart(HNR ~ Duration + F0.Start + F0.End + F0.Min + F0.Max + Formant.1 +
Formant.2,data=dftrain)

plot(HNRcart,xlim=c(0.75,5.33),ylim=c(-0.01,1.11))
text(HNRcart,use.n=TRUE)
dev.new()
rsq.rpart(HNRcart)
dev.new()
predHNR <-predict(HNRcart,dftest)
plot(dftest$HNR,predHNR,xlab=c("Actual HNR"),ylab=c("Predicted HNR"),main="CART
Predictions for TestData")
```

FIGURE 12.8
CART tree code for elephant call data.

The *plot*() function, together with the *text*() function, shows a drawing of the tree, together with predicted values for the training set. The function *rsq.rpart*() provides information about model quality (C_p, relative error), and the variables that were actually used in creating the tree. Figure 12.9 shows the output from the *rsq.rpart*() function. Figure 12.10 shows the tree diagram.

```
> rsq.rpart(HNRcart)

Regression tree:
rpart(formula = HNR ~ Duration + F0.Start + F0.End + F0.Min +
    F0.Max + Formant.1 + Formant.2, data = dftrain)

Variables actually used in tree construction:
[1] F0.Max    Formant.1

Root node error: 8350/323 = 25.851

n = 323

      CP        nsplit     rel error    xerror       xstd
1 0.834223      0          1.000000    1.004062    0.0816923
2 0.078678      1          0.165777    0.169753    0.0159373
3 0.037426      2          0.087099    0.091525    0.0072217
4 0.012581      3          0.049673    0.054851    0.0058374
5 0.010000      4          0.037092    0.043777    0.0052358
```

FIGURE 12.9
The rsq.rpart() output—building the tree.

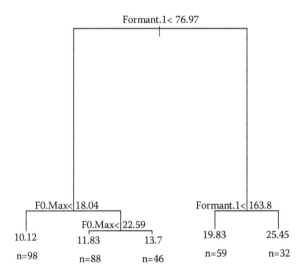

FIGURE 12.10
CART tree diagram.

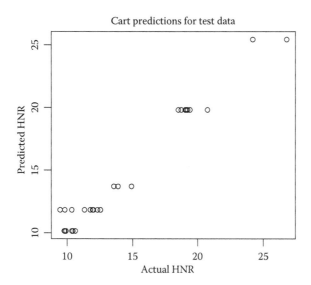

FIGURE 12.11
Predicted and actual HNR values for test data set.

The *predict()* function can be used to generate model predictions for a test set of data. Figure 12.11 shows the predicted values for the test data plotted against the actual observed HNR values.

Bayesian Model Averaging

Suppose there were several possible models that could all provide predictions of some response variable, but it was not clear which model provides the best prediction. The stepwise regression approach picks models based on minimizing the mean square error. Another possible approach is to compute a sort of weighted average of predictions from many models. The Bayesian Model Averaging (BMA) approach attempts to provide such a weighted average prediction. Hoeting et al. (1999) provide an excellent explanation of the BMA approach. The idea is fairly simple. If Y is the response variable of interest, suppose $\hat{Y} = f_i(X|Y)$ is a predicted value of the response with regressors X, data Y, using model f_i. If $i = 1, k$, then there would be k predictions given the data. Since models have random noise components, then predictions are in fact random variables, and as such have probability distributions/density functions. Now define

$$g_i\left(\hat{Y}|f_i, Y\right)$$

to be the density function of predictions generated from model f_i with data Y. The Bayesian part begins by postulating that there is a probability that model f_i is the "correct" model given data Y. Let this probability be symbolized as

$$p_i\left(f_i\middle|Y\right)$$

So, the weighted density function for the prediction would be

$$h\left(\hat{Y}\middle|Y\right)=\sum_{i=1}^{k}g_i\left(\hat{Y}\middle|fi,Y\right)p_i\left(fi\middle|Y\right)$$

In the non-Bayesian paradigm, a likelihood function for the data given the model might be formulated:

$$l_i\left(Y\middle|f_i\right)$$

Then, if, before any data are gathered, there is some a priori guess about how likely it is that each model is the "correct" model, symbolized by $r_i(f_i)$, then

$$p_j\left(f_j\middle|Y\right)=\frac{l_j\left(Y\middle|f_j\right)r_j(f_j)}{\sum_{i=1}^{k}l_i\left(Y\middle|f_i\right)r_i(f_i)}$$

This process involves defining a lot of models. There are the f_i, the g_i, the l_i, and r_i.

In practice, the models are generally of the same form, and the prior probabilities are assigned to the parameters of the model. So, in our elephant call example, the linear model

$$\text{HNR} = \beta_0 + \beta_1\text{Duration} + \beta_2\text{F0.Start} + \beta_3\text{F0.End} + \beta_4\text{F0.Min}$$
$$+ \beta_5\text{F0.Max} + \beta_6\text{Format.1} + \beta_7\text{Format.2} + \text{noise}$$

has eight parameters. The parameters would all be given prior "weights." The posterior weights would be computed based on the choice of the likelihood functions, which would again in practice all be from the same "family" (e.g., Gaussian, that is, normal). The function *bic.glm*() in package BMA will yield the posterior average coefficient values for each variable. These in turn can then be used to compute an average predicted value for each set of regressors.

The code for using BMA with the elephant call data is given in Figure 12.12. The output of *summary*() results for both an ordinary least squares (OLS) model with all seven regressors, and for the BMA model, are given in Figure 12.13.

```
setwd("/Users/scottpardo/SMFSBE/programs & data/")
df1 <-read.csv("20170301 Example 12.1 Elephant Calls.csv")
#
#
#  need to execute library(BMA) once during R session
library(BMA)
#
# VARIABLES:
# Call.ID
# CallType
# Duration
# F0.Start
# F0.End
# F0.Min
# F0.Max
# Formant.1
# Formant.2
# HNR
# testind
#
nbin <-nrow(df1)
#testind <-rbinom(n=nbin,size=1,prob=0.1) #currently testind is an
 input column
attach(df1)
fCall.ID <-factor(Call.ID)
levels(fCall.ID) <-
 c("let's.go","post.cop","greet","cont.call","cont.answ","bee","predato
 r")
df2 <-data.frame(cbind(df1,fCall.ID))
detach(df1)
attach(df2)
dftest <-subset(df2,testind==1)
dftrain <-subset(df2,testind==0)
fCall.ID <-factor(Call.ID)
#reg.df <-
 data.frame(cbind(Duration,F0.Start,F0.End,F0.Min,F0.Max,Formant.1,Form
 ant.2,HNR))
#
detach(df2)
attach(dftrain)
bma.model <-bic.glm(f=HNR~ Duration + F0.Start + F0.End + F0.Min +
 F0.Max + Formant.1 + Formant.2,data=dftrain,glm.family=gaussian())
linreg <-lm(HNR ~ Duration + F0.Start + F0.End + F0.Min + F0.Max +
 Formant.1 + Formant.2,data=dftrain)
postcoeff <-bma.model$postmean
HNRpred <-postcoeff[1] + postcoeff[2]*Duration +
 postcoeff[3]*F0.Start + postcoeff[4]*F0.End + postcoeff[5]*F0.Min +
 postcoeff[6]*F0.Max + postcoeff[7]*Formant.1 + postcoeff[8]*Formant.2;
# Make some plots:
plot(x=HNR,y=HNRpred,pch=1,ylab="HNR")
points(x=HNR,y=linreg$fitted.values,pch=2)
points(x=HNR,y=HNR,pch=3)
legend(x=10.5,y=25.0,legend=c("BMA","OLS","Observed HNR"),pch=c(1,2,3))
#
#
summary(linreg)
summary(bma.model)
```

FIGURE 12.12
Bayesian model averaging code for elephant call data.

```
Call:
lm(formula = HNR ~ Duration + F0.Start + F0.End + F0.Min + F0.Max +
   Formant.1 + Formant.2, data = dftrain)

Residuals:
   Min       1Q    Median       3Q      Max
-0.54808 -0.15560  0.00631  0.14908  0.56458

Coefficients:
              Estimate   Std. Error   t value   Pr(>|t|)
(Intercept)   0.1042071   0.1136467    0.917     0.360
Duration      0.0021198   0.0043661    0.486     0.628
F0.Start      0.0034224   0.0111502    0.307     0.759
F0.End        0.0045839   0.0083710    0.548     0.584
F0.Min        0.2320318   0.0100651   23.053    <2e-16 ***
F0.Max        0.2515426   0.0036079   69.719    <2e-16 ***
Formant.1     0.0997243   0.0005533  180.227    <2e-16 ***
Formant.2    -0.0099631   0.0002978  -33.460    <2e-16 ***
---
Signif. codes:  0 '***' 0.001 '**' 0.01 '*' 0.05 '.'0.1 ' ' 1

Residual standard error: 0.2101 on 315 degrees of freedom
Multiple R-squared:  0.9983,Adjusted R-squared:  0.9983
F-statistic: 2.697e+04 on 7 and 315 DF,  p-value: < 2.2e-16

> summary(bma.model)

Call:
bic.glm.formula(f = HNR ~ Duration + F0.Start + F0.End + F0.Min + F0.Max + Formant.1 +
    Formant.2, data = dftrain, glm.family = gaussian())


  4  models were selected
  Best  4  models (cumulative posterior probability =  1 ):

                p!=0        EV          S       model 1     model 2     model 3
Intercept     100.0    0.1596269   0.0915261   1.628e-01   1.304e-01   1.433e-01
Duration.x      4.7    0.0000315   0.0008769       .           .           .
F0.Start.x      5.4    0.0002549   0.0023737       .           .       4.761e-03
F0.End.x        5.9    0.0002754   0.0020186       .       4.697e-03       .
F0.Min.x      100.0    0.2371131   0.0065       2.375e-01   2.346e-01   2.336e-01
F0.Max.x      100.0    0.2526709   0.0032524   2.527e-01   2.523e-01   2.523e-01
Formant.1.x   100.0    0.0997400   0.0005490   9.974e-02   9.971e-02   9.972e-02
Formant.2.x   100.0   -0.0100399   0.0002788  -1.004e-02  -1.001e-02  -1.001e-02

                model 4
Intercept     1.587e-01
Duration.x    6.643e-04
F0.Start.x        .
F0.End.x          .
F0.Min.x      2.375e-01
F0.Max.x      2.526e-01
Formant.1.x   9.975e-02
Formant.2.x  -1.004e-02
            model 1      model 2      model 3      model 4
nVar           4            5            5            5
BIC      -1.522e+03    -1.516e+03   -1.516e+03   -1.516e+03
post prob   0.840        0.059        0.054        0.047
```

FIGURE 12.13
BMA summary output for elephant call data.

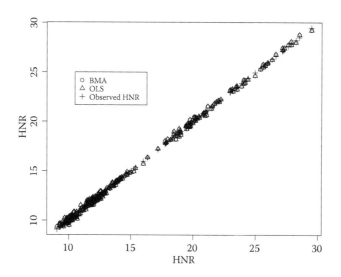

FIGURE 12.14
Predicted HNR plotted against observed.

Figure 12.14 shows predicted HNR values plotted against observed, for both OLS and BMA models. The column in the BMA summary called "EV" is the Bayesian posterior expected value for each of the coefficients.

Theoretical Aspects

The techniques described in this chapter rely on algorithms. In a sort of pseudo-code language, a algorithm might be symbolized as in Figure 12.15.

The line "Update x" could involve very complicated procedures, decisions and computations, depending on the nature of the algorithm. The idea though is that all algorithms require some initial state for input variables or parameters, an objective function, a means of updating the inputs in a way that results in new values for the objective function, and a rule for deciding that the inputs have been updated sufficiently.

Figure 12.16 is an illustration of African elephants.

```
Initialize input vector x;

Define delta, the required change in the objective function;

Define objective function, f(x);

Define updating rule for input vector x;

Define stopping rule in terms of f(x);

Until stopping rule is achieved, do:

 x old<-x;

 Update x;

 x new<-Updated x;

 Compute f(xnew)

 If | f(xnew) -f(xold)| ≤ delta

  Stopping rule achieved;

 x <-x new;

 loop
```

FIGURE 12.15
An algorithm description.

FIGURE 12.16
African elephants (*Loxodonta africana*).

Key Points for Chapter 12

- In some cases, especially when many potential explanatory variables/regressors are present, determining the best predictive model may be challenging.

- Several techniques, sometimes as referred to as "machine learning methods," were presented.

- These methods provide a predictive model without necessarily understanding the relationship between the response and the regressors.

- The methods are "algorithmic" in nature, and only require fairly general input from the analyst.

- While they can provide good predictions of the response, model-building algorithms do not in general provide much insight into the physical, biological, or social relationships between the regressors and the response.

Exercises and Questions

1. Plot predictions of HNR against observed HNR. Compare predictions of HNR made by each of the techniques of this chapter. Do you find one set of predictions better than any of the others?

2. Compute the residuals for each prediction method, using the "dftraining" data frame. Then compute the mean and standard deviation for those residuals.

3. Perform a one-way ANOVA on the residuals for all the models using the elephant call data (dftrain). Is there a significant difference?

4. The bic.glm() function has a parameter called prior.param, that is a vector of "weights" the analyst can give to each potential regressor; it defaults to 0.5 for each variable. Try executing the model script with prior weights prior.param=c(0.9,0.9,0.9,0.1,0.1,0.1,0.1); Do you obtain different average coefficients? Did the algorithm select a different number of intermediate models?

13

Time-to-Event

General Ideas

A very common type of data in behavioral ecology is what is known as time-to-event data. An event is something that occurs in an instant (as opposed to being measured over a span of time), and time-to-event refers to the amount of time elapsed between a predetermined "start" point and the occurrence of the event in question. In behavioral ecology studies, the event is usually a particular behavior performed by the subject(s). A few examples of time-to-event data include the amount of time until an animal responds to a playback or a predator mount, the amount of time until the start of an annual migration, and the amount of time until a juvenile disperses from its natal territory.

The term *survival* refers to the probability that an event will occur within t time units after the start point. The event in question could be death, in which case "survival" is meant literally, but this need not be the case. One might consider literal "survival," that is, probability that an individual lives to a certain age, as a special case of probabilities that some specific type of event will occur no sooner than t units from some initial reference time. The point is to derive a model by which the probability of interest can be predicted. In general, a curve showing the probability of surviving (y-axis) plotted against the time-to-event will be computed. Such a curve is known as a survival curve.

In many cases, some of the individuals in a study may be observed for a period of time without ever observing the event of interest. Such observations are referred to as "censored" (Lee, 1992). In this chapter we present a method for incorporating such censored observations called the Kaplan–Meier (K-M) product limit approach. In addition to estimating survival, sometimes the researcher desires to compare survival between groups or treatments. We discuss two methods for comparing survival curves, the Logrank test (or a variant called the Peto and Peto test), and the Cox proportional hazard model. The Logrank test is a hypothesis-testing method, whereas the Cox proportional hazard is a form of generalized linear model.

Examples with R Code

Male fruit flies *(Drosophila melanogaster)* have an organ called an accessory gland, which secretes certain proteins into their seminal fluid. These proteins increase the female's egg-laying rate and decrease her receptivity to other males, thus increasing the male's own paternity. Chapman et al. (1995) studied the effects of male accessory gland proteins (Acps) on the survival of female *D. melanogaster*. As an example inspired by this study, suppose that 30 female flies were observed for 25 days after exposure to Acps. Of the 30, five of them were still alive after the 25 days. Thus, those five would be considered censored observations. The remaining 25 had different survival times.

Figure 13.1 shows R code (and output) for computing the K-M survival statistics with function *survfit()* in package survival, and the computing of a predictive model. The predictive model can be used to interpolate to survival times other than those actually observed in the dataset.

Figure 13.2 shows a plot of the K-M estimates and the predicted survival probabilities from the fitted model.

Comparison of Survival Curves

The R function, *survdiff()*, in package survival, implements the LogRank or Peto and Peto test. It allows the user to choose the scoring method with an input parameter rho. If rho is set to zero (the default), the scores are computed via the LogRank approach. If rho is set equal to one, then the Peto and Peto method is employed. A formula parameter allows for a linear model specification. The test statistic will be computed for the linear model overall, and not for each term in the model. The *survdiff()* function also allows for a "stratification" variable, using the *strata()* function, that accounts for "blocking" or nuisance effects (such as different groups of individuals).

Consider the following example based on the work of Webster et al. (2011). African wild dogs *(Lycaon pictus)* are sympatric with several other species of large carnivores throughout their range, including lions *(Panthera leo)* and spotted hyenas *(Crocuta crocuta)*. Both lions and hyenas directly compete with wild dogs for food, and may kill wild dogs as well. However, lions are much more likely than hyenas to kill adult wild dogs. Webster et al. (2011) investigated whether African wild dogs perceive lions and hyenas as presenting different levels of threat. In addition, because habitat visibility affects the likelihood of being ambushed by other predators, the researchers also investigated whether the different levels of visibility afforded by different habitat types affected the wild dogs' reaction to hearing lions and hyenas nearby. The researchers played back lion roars and hyena whoops to groups of African wild dogs in three different habitat types: Open, intermediate, and dense. They then measured the elapsed time from the moment that the first dog raised its head in reaction to the playback until the moment when

```
setwd("C:\\Users\\SMFSBE\\Statistical Data & Programs")
rm(T2E,Censor,surv.obj,surv.model)
df1 <-read.csv("20170502 Example 13.1 Fruit Flies and Acps.csv")
#
#
#   Need library(survival) for this code to work
#   only needs to run once in a session
library(survival)
#

# VARIABLES:
#
# Censor
# T2E
#
attach(df1)

surv.obj <-Surv(time=T2E,event=Censor)


surv.model <-survfit(formula = surv.obj ~ 1,data=df1)

summary(surv.model)
#
#     Selected attributes of object surv.model:
#     surv.model$n:  total sample size
#     surv.model$time: vector of survival times
#     surv.model$surv: vector of survival probability estimates
#     surv.model$type: type of censoring
#     surv.model$std.err: vector of survival probability estimate standard errors
#     surv.model$upper: vector of survival probability estimate upper confidence limits
#     surv.model$lower: vector of survival probability estimate lower confidence limits
#     surv.model$conf.type: transformation used to compute confidence limits
#     surv.model$conf.int: confidence level
#
#
# Now fit a predictive model for survival time
#
s.time <-surv.model$time
s.time.sq <-s.time**2
Ht <- -log(surv.model$surv)
Ht.model <-lm(Ht ~ s.time + s.time.sq)
#

#   Check quality of model fit:
#
summary(Ht.model)
B0 <-Ht.model$coefficients[1]
B1 <-Ht.model$coefficients[2]
B2 <-Ht.model$coefficients[3]
#
#   Compute the predicted survival probabilities
#
pred.surv <-exp(-(B0 + B1*s.time + B2*s.time.sq))

#plot(x=surv.model$time[factor(Treat)==1],y=surv.model$surv[factor(Treat)==1],pch=1,col="bl
ue",type="p")
#dev.new()
#plot(x=surv.model$time[factor(Treat)==2],y=surv.model$surv[factor(Treat)==2],pch=2,col="re
d",type="p")
plot(x=surv.model$time,y=surv.model$surv,main="Female Drosophila Survival",xlab="Survival
Time (days)",ylab="Probability",pch=3,col="black",type="p")
points(x=surv.model$time,y=pred.surv,pch=1,col="blue",type="p")
legend(x=17.0,y=0.80,legend=c("Kaplan-Meier Survival","Model
Predictions"),col=c("black","blue"),pch=c(3,1))
```

FIGURE 13.1
Survival curve code. *(Continued)*

```

> setwd("C:\\Users\\SMFSBE\\Statistical Data & Programs")
> rm(T2E,Censor,surv.obj,surv.model)
Warning messages:
1:In rm(T2E, Censor, surv.obj, surv.model) : object 'T2E' not found
2:In rm(T2E, Censor, surv.obj, surv.model) : object 'Censor' not found
> df1 <-read.csv("20170502 Example 13.1 Fruit Flies and Acps.csv")
> #
> #
> #  Need library(survival) for this code to work
> #   only needs to run once in a session
> library(survival)
> #
>
> # VARIABLES:
> #
> # Censor
> # T2E
> #
> attach(df1)
>
> surv.model <-survfit(formula = surv.obj ~ 1,data=df1)
>
> summary(surv.model)
Call: survfit(formula = surv.obj ~ 1, data = df1)

 time n.risk n.event survival std.err lower 95% CI upper 95% CI
  9.3     30       1    0.967  0.0328       0.9045        1.000
  9.7     29       1    0.933  0.0455       0.8482        1.000
  9.8     28       1    0.900  0.0548       0.7988        1.000
 11.0     27       1    0.867  0.0621       0.7532        0.997
 11.2     26       1    0.833  0.0680       0.7101        0.978
 11.4     25       2    0.767  0.0772       0.6293        0.934
 11.7     23       1    0.733  0.0807       0.5910        0.910
 13.4     22       1    0.700  0.0837       0.5538        0.885
 13.7     21       1    0.667  0.0861       0.5176        0.859
 14.6     20       1    0.633  0.0880       0.4824        0.832
 15.1     19       2    0.567  0.0905       0.4144        0.775
 15.4     17       1    0.533  0.0911       0.3816        0.745
 16.4     16       2    0.467  0.0911       0.3183        0.684
 16.7     14       1    0.433  0.0905       0.2878        0.652
 18.1     13       1    0.400  0.0894       0.2581        0.620
 18.7     12       1    0.367  0.0880       0.2291        0.587
 18.8     11       1    0.333  0.0861       0.2010        0.553
 20.0     10       1    0.300  0.0837       0.1737        0.518
 20.5      9       1    0.267  0.0807       0.1473        0.483
 22.0      8       1    0.233  0.0772       0.1220        0.446
 22.5      7       1    0.200  0.0730       0.0978        0.409
 23.0      6       1    0.167  0.0680       0.0749        0.371
> #
> #   Selected attributes of object surv.model:
> #   surv.model$n:  total sample size
> #   surv.model$time: vector of survival times
> #   surv.model$surv:  vector of survival probability estimates
> #   surv.model$type: type of censoring
> #   surv.model$std.err: vector of survival probability estimate standard errors
> #   surv.model$upper: vector of survival probability estimate upper confidence limits
> #   surv.model$lower: vector of survival probability estimate lower confidence limits
> #   surv.model$conf.type: transformation used to compute confidence limits
> #   surv.model$conf.int: cobnfidence level
> #
> #
> # Now fit a predictive model for survival time
> #
> s.time <-surv.model$time
> s.time.sq <-s.time**2
> Ht <--log(surv.model$surv)
> Ht.model <-lm(Ht ~ s.time + s.time.sq)
> #
> #  Check quality of model fit:
> #
> summary(Ht.model)
```

FIGURE 13.1 (CONTINUED)
Survival curve code.   *(Continued)*

```
Call:
lm(formula = Ht ~ s.time + s.time.sq)

Residuals:
    Min        1Q     Median        3Q        Max
-0.147297 -0.034233 -0.002464  0.048145  0.159619

Coefficients:
            Estimate Std. Error t value Pr(>|t|)
(Intercept) -0.5683520  0.1806693  -3.146  0.00509 **
s.time       0.0425119  0.0230169   1.847  0.07960 .
s.time.sq    0.0023114  0.0006899   3.350  0.00319 **
---
Signif. codes:  0 '***' 0.001 '**' 0.01 '*' 0.05 '.' 0.1 ' ' 1

Residual standard error: 0.06505 on 20 degrees of freedom
Multiple R-squared:  0.988,      Adjusted R-squared:  0.9868
F-statistic: 826.3 on 2 and 20 DF,  p-value: < 2.2e-16
> B0 <-Ht.model$coefficients[1]
> B1 <-Ht.model$coefficients[2]
> B2 <-Ht.model$coefficients[3]
> #
> #  Compute the predicted survival probabilities
> #
> pred.surv <-exp(-(B0 + B1*s.time + B2*s.time.sq))
>
>
#plot(x=surv.model$time[factor(Treat)==1],y=surv.model$surv[factor(Treat)==1],pch=1,col="bl
ue",type="p")
> #dev.new()
>
#plot(x=surv.model$time[factor(Treat)==2],y=surv.model$surv[factor(Treat)==2],pch=2,col="re
d",type="p")
> plot(x=surv.model$time,y=surv.model$surv,pch=3,col="black",type="p")
> points(x=surv.model$time,y=pred.surv,pch=1,col="blue",type="p")
> legend(x=17.0,y=0.80,legend=c("Kaplan-Meier Survival","Model
Predictions"),col=c("black","blue"),pch=c(3,1))
>
```

FIGURE 13.1 (CONTINUED)
Survival curve code.

the last dog departed the area (latency to pack retreat). In our example, we will only focus on lion roars, and the latency (time-to-event) distributions as they are affected by habitat type. The example has 16 different wild dog packs, each observed in each of the three habitat types (open, intermediate, and dense). The censoring time is 60 minutes. The question is whether the latency time distribution is affected by the type of habitat. Figure 13.3 shows the R code using the simulated wild dog data, and the associated R console output. The R function *coxph()*, also in package survival, performs the computations, where the response variable is a *Surv()* object. Figure 13.4 shows survival curves for each of the habitats.

Theoretical Aspects

To begin, we borrow from elementary chemical kinetics (Whitten et al., 2004), and consider a first-order system, that is, a function of time that can be

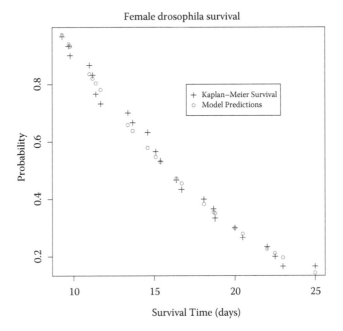

FIGURE 13.2
Survival probabilities for female fruit flies.

expressed as a first-order differential equation. Suppose we can measure a response variable that indicates the degree to which an individual has some characteristic, and further suppose that this response variable is decreasing over time. Let $Y(t)$ be the response variable observed at time t. A first-order differential equation model that describes the change in this response over time is

$$\frac{dY(t)}{dt} = -\lambda Y(t)$$

With initial condition $Y(0) = y_0$, the solution to the equation is

$$Y(t) = y_0 e^{-\lambda t}$$

The more chemically minded among us may recognize this as the equation governing first-order chemical reaction kinetics (Whitten et al., 2004), or the equation for radioactive decay (Rutherford, 1900). The parameter λ is called the rate parameter (as in the rate of reaction). If we presume that the response variable $Y(t)$ is a decreasing function of time, as in the case of analyte concentrations in chemical reactions, then it is a maximum at $t = 0$, and

```
setwd("C:\\Users\\SMFSBE\\Statistical Data & Programs")
rm(T2E,Censor,surv.obj,surv.model)
df1 <-read.csv("20170510 Example 13.2 Wild Dogs and Lions.csv")
#
#
#  Need library(survival) for this code to work
#  only needs to run once in a session
library(survival)
#

# VARIABLES:
#
# Group
# Habitat
# Latency
# Censoring
#
attach(df1)

lat.obj <-Surv(time=Latency,event=Censoring)
survdiff(formula=lat.obj ~ strata(Group) + Habitat,rho=0)#LogRank Scores
survdiff(formula=lat.obj ~ strata(Group) + Habitat,rho=1)#Peto and Peto Scores

cox.model <-coxph(formula = lat.obj ~ Group + Habitat,data=df1)
summary(cox.model)

surv.model <-survfit(formula = lat.obj ~ Habitat,data=df1)

summary(surv.model)
plot(surv.model,lty=c(1,2,3),pch=c(1,2,3),xlab="Latency
(minutes)",ylab="S(t)",col=c("black","blue","red"))
legend(x=30.0,y=0.70,legend=c("Open","Intermediate","Dense"),lty=c(1,2,3),col=c("black","blu
e","red"))
#
#    Selected attributes of object surv.model:
#    surv.model$n:   total sample size
#    surv.model$time: vector of survival times
#    surv.model$surv:  vector of survival probability estimates
#    surv.model$type: type of censoring
#    surv.model$std.err: vector of survival probability estimate standard errors
#    surv.model$upper: vector of survival probability estimate upper confidence limits
#    surv.model$lower: vector of survival probability estimate lower confidence limits
#    surv.model$conf.type: transformation used to compute confidence limits
#    surv.model$conf.int: confidence level
#
#
#
OUTPUT

>
> lat.obj <-Surv(time=Latency,event=Censoring)
> survdiff(formula=lat.obj ~ strata(Group) + Habitat,rho=0)#LogRank Scores
Call:
survdiff(formula = lat.obj ~ strata(Group) + Habitat, rho = 0)

            N Observed Expected (O-E)^2/E (O-E)^2/V
Habitat=1 16      15    24.33      3.580     12.35
Habitat=2 16      14    11.83      0.397      0.69
Habitat=3 16      13     5.83      8.805     13.50

Chisq= 18.6  on 2 degrees of freedom, p= 9.21e-05
> survdiff(formula=lat.obj ~ strata(Group) + Habitat,rho=1)#Peto and Peto Scores
Call:
survdiff(formula = lat.obj ~ strata(Group) + Habitat, rho = 1)

            N Observed Expected (O-E)^2/E (O-E)^2/V
Habitat=1 16    6.67    14.00      3.84      10.5
Habitat=2 16    9.67     9.67      0.00       0.0
Habitat=3 16   13.00     5.67      9.49      14.7

Chisq= 17.7  on 2 degrees of freedom, p= 0.000142
>
> cox.model <-coxph(formula = lat.obj ~ Group + Habitat,data=df1)
> summary(cox.model)
Call:
coxph(formula = lat.obj ~ Group + Habitat, data = df1)

 n= 48, number of events= 42

             coef  exp(coef)  se(coef)      z   Pr(>|z|)
Group   -0.008097   0.991936  0.035958  -0.225   0.8218
Habitat  0.458638   1.581918  0.191614   2.394   0.0167 *
---
Signif. codes:  0 '***' 0.001 '**' 0.01 '*' 0.05 '.' 0.1 ' ' 1

          exp(coef)exp(-coef) lower .95 upper .95
Group        0.9919    1.0081    0.9244     1.064
Habitat      1.5819    0.6321    1.0866     2.303
```

FIGURE 13.3
R code and output for wild dog example. *(Continued)*

```
Concordance= 0.737   (se = 0.052 )
Rsquare= 0.111    (max possible= 0.996 )
Likelihood ratio test= 5.66  on 2 df,   p=0.05891
Wald test           = 5.82  on 2 df,   p=0.05434
Score (logrank) test = 6.01  on 2 df,   p=0.04948


>
> surv.model <-survfit(formula = lat.obj ~ Habitat,data=df1)
>
> summary(surv.model)
Call: survfit(formula = lat.obj ~ Habitat, data = df1)
```

```
        Habitat=1
  time  n.risk  n.event survival std.err lower 95% CI upper 95% CI
   7.8    16      1     0.9375   0.0605    0.82609      1.000
   8.4    15      1     0.8750   0.0827    0.72707      1.000
   8.8    14      2     0.7500   0.1083    0.56520      0.995
   9.0    12      2     0.6250   0.1210    0.42761      0.914
   9.2    10      1     0.5625   0.1240    0.36513      0.867
   9.3     9      1     0.5000   0.1250    0.30632      0.816
   9.6     8      1     0.4375   0.1240    0.25101      0.763
   9.9     7      1     0.3750   0.1210    0.19921      0.706
  10.2     6      1     0.3125   0.1159    0.15108      0.646
  10.3     5      1     0.2500   0.1083    0.10699      0.584
  11.1     4      1     0.1875   0.0976    0.06761      0.520
  12.7     3      1     0.1250   0.0827    0.03419      0.457
  13.6     2      1     0.0625   0.0605    0.00937      0.417
```

```
        Habitat=2
  time  n.risk  n.event survival std.err lower 95% CI upper 95% CI
   3.9    16      1      0.938   0.0605    0.8261       1.000
   4.0    15      1      0.875   0.0827    0.7271       1.000
   4.2    14      1      0.812   0.0976    0.6421       1.000
   4.4    13      1      0.750   0.1083    0.5652       0.995
   4.5    12      1      0.688   0.1159    0.4941       0.957
   4.7    11      2      0.563   0.1240    0.3651       0.867
   4.8     9      1      0.500   0.1250    0.3063       0.816
   5.1     8      1      0.438   0.1240    0.2510       0.763
   5.3     7      1      0.375   0.1210    0.1992       0.706
   5.4     6      1      0.312   0.1159    0.1511       0.646
   5.5     5      1      0.250   0.1083    0.1070       0.584
   5.7     4      1      0.188   0.0976    0.0676       0.520
   6.1     3      1      0.125   0.0827    0.0342       0.457
```

```
        Habitat=3
  time  n.risk  n.event survival std.err lower 95% CI upper 95% CI
   0.6    16      1      0.938   0.0605    0.8261       1.000
   0.7    15      3      0.750   0.1083    0.5652       0.995
   0.8    12      4      0.500   0.1250    0.3063       0.816
   0.9     8      1      0.438   0.1240    0.2510       0.763
   1.1     7      2      0.312   0.1159    0.1511       0.646
   1.2     5      1      0.250   0.1083    0.1070       0.584
   1.6     4      1      0.188   0.0976    0.0676       0.520
```

FIGURE 13.3 (CONTINUED)
R code and output for wild dog example.

its maximum value is y_0. Thus we can express the response as a proportion of its initial value, namely:

$$\frac{Y(t)}{y_0} = e^{-\lambda t}$$

This proportion can be thought of as a probability; namely, the probability that the value of the response is not zero after t time units. So, if we replace

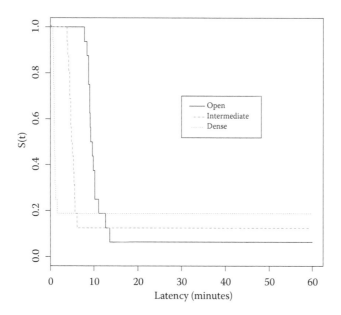

FIGURE 13.4
Survival (latency) probability curves for wild dog example by habitat.

$Y(t)$ with a new variable, namely T = the time at which $Y(t)$ is zero, then we can express the value of $Y(t)$ as a proportion of its initial value with $Pr\{T \geq t\}$, or

$$Pr\{T \geq t\} = e^{-\lambda t}$$

It turns out that this is the complement of the cumulative distribution function of an exponential random variable. That is,

$$Pr\{T \leq t\} = 1 - Pr\{T \geq t\} = 1 - e^{-\lambda t}$$

In the language of Survival, the parameter λ is called the *hazard rate*. The exponential time-to-failure variable is characterized by a constant hazard rate. In other words, the rate of occurrence for the event of interest (i.e., average number of events per unit time) could potentially change with time t. If $h(t)$ represents the hazard rate, then for the exponential time to failure variable,

$$h(t) = \lambda \forall t$$

That is, in the case of exponentially distributed time-to-event variables, the hazard rate does not change as time goes on. Furthermore, a convenient feature of the exponential model is that the average time to event is the recip-rocal of the (constant) hazard rate.

The exponential time-to-failure variable is related to a discrete random variable with a Poisson distribution. The variable, X, is the number of fail-ures (or events) within a fixed length of time. It has a Poisson distribution if

$$Pr\{X = x \mid \lambda, t\} = \frac{(\lambda t)^x e^{-\lambda t}}{x!}$$

The value of t is a fixed length of time, and λ is the (constant) failure rate.

Suppose that the hazard rate actually changes with time. For example, the death rate could be fairly high initially (infant mortality), then drop after some age to a constant, and then climb back up after the individual reaches a certain age. Such a hazard rate function is sometimes called a "bathtub" curve, as illustrated in Figure 13.5.

The failure rate function is also called the hazard rate function.

To generalize our initial first-order differential equation model for the response variable $Y(t)$, we could replace the constant λ with $h(t)$:

$$\frac{dY(t)}{dt} = -h(t)Y(t)$$

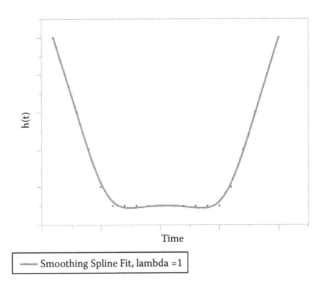

Time

───── Smoothing Spline Fit, lambda =1

FIGURE 13.5
"Bathtub" failure rate curve.

Again using the initial condition $Y(0) = y_0$, the solution is

$$Y(t) = y_0 e^{-\int_0^t h(\tau) d\tau}$$

Of course, if $h(t) = \lambda$ (a constant), then:

$$\int_0^t h(\tau) d\tau = \lambda t$$

The function:

$$H(t) = \int_0^t h(\tau) d\tau$$

is called the cumulative hazard rate function for our new random variable T, time to event. So, in general, using our slightly generalized first order kinetics model,

$$Pr\{T \geq t\} = e^{-\int_0^t h(\tau) d\tau} = e^{-H(t)}$$

If we define the survival function to be

$$S(t) = Pr\{T \geq t\} = e^{-\int_0^t h(\tau) d\tau} = e^{-H(t)}$$

Then the reliability function can be thought of as a curve in time. We also have some potentially useful relationships:

$$-\ln S(t) = H(t)$$

$$h(t) = \frac{dH(t)}{dt}$$

Another related pair of functions is

$$F(t) = 1 - S(t) = 1 - e^{-H(t)}$$

$$G(t) = -\ln F(t)$$

$F(t)$ is the cumulative distribution function for time-to-event.
Thus we have the relationship:

$$S(t) = 1 - e^{-G(t)} = e^{-H(t)}$$

Obtaining an Empirical Survival Model

Generally the researcher does not know the specific form of $S(t)$, $H(t)$, or $h(t)$. She or he could perform an experiment to obtain a polynomial approximation. In this experiment, n individuals (or groups, if group is the experimental and observational unit) will be observed from the pre-defined starting point until the event of interest occurs. The elapsed time from start to end will be recorded. Suppose the n times-to-event are ordered from shortest to longest. Call these times t_1, t_2, ..., t_k, ..., t_n. These times are referred to as "order statistics" (Conover, 1980). Compute the empirical reliability function:

$$\hat{S}(t_k) = 1 - \frac{k}{n} = \frac{n-k}{n} \quad k = 1, n$$

or the empirical cumulative distribution function:

$$\hat{F}(t_k) = 1 - \hat{S}(t_k) = \frac{k}{n} \quad k = 1, n$$

For a lack of a better term, we will call this formula, or estimator, the empirical maximum likelihood (EML) estimator. Now suppose that the cumulative hazard rate function, $H(t)$, can be approximated by a low order polynomial in t, for example:

$$H(t) = \beta_0 + \beta_1 t + \beta_2 t^2$$

Given the relation:

$$-\ln S(t) = H(t)$$

It may be advantageous to approximate $G(t)$ as a second-order polynomial, that is:

$$G(t) = \beta_0 + \beta_1 t + \beta_2 t^2$$

The researcher can now obtain an estimate of the parameters β_k, and thus an approximation formula for $R(t)$, via least squares. That is, the estimated approximation formula would be

$$\tilde{S}(t) = 1 - \exp(-G(t)) = 1 - \exp(-\hat{\beta}_0 - \hat{\beta}_1 t - \hat{\beta}_2 t^2)$$

The $\hat{\beta}_i$ are the least squares estimates of the $H(t)$ approximation formula parameters. The approximation formula could be used to interpolate values of $S(t)$.

Censored Time-to-Failure

Sometimes limits are placed on the length of time over which experimental units will be observed, or on the number of individuals (out of a sample of size n) for which the event of interest occurs before the observations are curtailed. Such restrictions are called censoring. Stopping the observations at a fixed time is referred to as Type I censoring, and stopping after a fixed number of events occur is called Type II censoring (Mann et al., 1974). The question is how to estimate survival in the face of such censoring. We will address Type I censoring first.

Suppose the researcher did an experiment with a sample of n experimental units, and stopped the test after T_{max} time units. Out of the n units on test, only $r < n$ were observed to have the event of interest occur. The event for the remaining $n - r$ units had not been observed at time T_{max}. Suppose further that the r times-to-event are sequenced from shortest to longest, and that t_1, $t_2, t_3, \ldots, t_r$ represent those order statistics. So the empirical survival function could be represented as it was when there was no censoring:

$$\hat{S}(t_k) = 1 - \frac{k}{n} = \frac{n - k}{n} \quad k = 1, r$$

This is the EML with right-censoring (EMLC). It is called "right-censoring" since the possible event of interest could occur some time in the future, to the "right" of the time at which the individual was no longer being observed. The researcher could treat this computed empirical survival function in the exact same way he or she did when there was no censoring. There is another method for computing an empirical survival function at times other than those which were explicitly observed event times. The method is due to Kaplan and Meier (1958), and is described in Lee (1992). The formula for the Kaplan–Meier (K-M) estimator is

$$\hat{S}(t) = \prod_{t_k \le t} \frac{n - k}{n - k + 1}$$

where t_k are the failure times for uncensored observations. At $t = t_k$, this formula can be written as

$$\hat{S}(t_k) = \hat{S}(t_{k-1}) \frac{n-k}{n-k+1}$$

The K-M estimator is particularly useful if censoring can occur even though the time has not exceeded T_{max}. The event of interest may occur during observation, but due to some cause other than the particular stimulus/response of interest. For example, if the event of interest is the animal leaving a certain area after some attractant is introduced, the subject may leave due to some uncontrollable stimulus, such as a thunder clap or appearance of a predator.

The variance, and thus standard error, of the K-M survival estimate, can be approximated (Lee, 1992) by:

$$V\left(\hat{S}(t_k)\right) \approx \hat{S}^2(t_k) \sum_{i=1}^{k} \frac{1}{(n-i)(n-i+1)}$$

The approximate standard error is the square root of this variance approximation.

Using either the EML, EMLC, or K-M estimators, the highest value of the empirical survival function is at time t_1, and it is $(n-1)/n$. The lowest survival value would be at t_r. If $r = n$, then EML = EMLC, and at t_r, $S(t_r) = 0$. For the K-M estimator, this is not the case if there are any intermediate censored failures (i.e., if the event is censored before T_{max} due to some stimulus other than the one(s) of interest).

Comparison of Survival Distributions

We will discuss two methods for comparing the effects of treatments on the time-to-event, or survival, distributions.

Mantel–Cox LogRank and Peto and Peto Procedures

The derivation and explanation of this test is given by Peto and Peto (1972). The idea is to first sort all survival times, censored or not, from smallest to largest. Then each observation gets a score, which is somewhat complicated to compute. First, for each *unique* time t (multiple observations with the same time are only counted once), compute the number of observations with survival time, t (call it $s(t)$), and the number of observations with time t or greater (call it $r(t)$). Then, for each individual, compute the sum:

$$-e(t) = -\sum_{t_j \le t} \frac{s(t_j)}{r(t_j)}$$

Most of the time, the $s(t_j)$ will equal 1. The score for the individual is $w_i = 1 - e(t_i)$. If the observation is censored at time T, then its score is simply $-e(T)$. It turns out that the score $-e(t)$ is the log of the survival probability at time t. Thus, the test is named "LogRank." The test statistic is an average, for each group or treatment, of squared scores of the individuals in the group, divided by the average squared scores over all observations, regardless of the group.

Peto and Peto (1972) also suggested an alternative score for each observation. If the event times are sorted from smallest to largest, and $\hat{S}(t_i)$ is the survival probability for the ith observation (sorted by the survival times regardless of which group or treatment, or whether they are "complete" event times or censored, then the score for the individual is

$$w_i = \begin{cases} S(t_i) + S(t_{i-1}) - 1, & i \ge 2 \\ S(t_i) - 1, & i = 1, \text{ or if } t_i \text{ is a censored time} \end{cases}$$

Cox Proportional Hazard Model

A generalized linear model can also be used to evaluate the effects of various factors on survival distributions. Recall that every survival probability function has an associated hazard function, $h(t)$, with the relationship:

$$S(t) = Pr\{T \ge t\} = e^{-\int_0^t h(\tau)d\tau} = e^{-H(t)}$$

Therefore, solving for the hazard function yields:

$$h(t) = \frac{d}{dt} - \ln S(i) = -\frac{S'(t)}{S(t)}$$

S' represents the derivative of S. The function $S'(t)$ can be approximated as

$$\hat{S}'(t_i) = \frac{\hat{S}(t_i) - \hat{S}(t_{i-1})}{t_i - t_{i-1}}$$

The Cox proportional hazard model presumes that there is some "base" hazard function and that the effective hazard function is related to the base through the existence of non-time related variables or factors. In particular, if h_0 is the base hazard, then the effective hazard function is

$$h_e(t) = h_0(t) f(x \mid \beta)$$

In other words, the effective hazard function is proportional to the base hazard function, and the proportionality is dependent on some function, $f(x \mid \beta)$ of the regressor/predictor/factor variables. In general, x is a vector of variables, and β is a vector of unknown parameters that need to be estimated. The function $f(x \mid \beta)$ could have any sort of form, but certain forms make the parameter estimation more tractable. Perhaps the most common form is exponential:

$$f\left(x \mid \beta\right) = \exp\left(\beta^T x\right)$$

Given the survival probability estimates for each time, $\hat{S}(t_i)$, an estimate of the effective hazard function can be computed:

$$\hat{h}_e(t_i) = -\frac{\hat{S}'(t_i)}{\hat{S}(t_i)}$$

The linearized model is

$$\ln\left(\hat{h}_e(t)\right) = \ln\left(h_0(t)\right) + \ln\left(f\left(x \mid \beta\right)\right) = \ln\left(h_0\left(t\right)\right) + \beta^T x$$

If it is assumed that the base survival function is exponential, then $h_0(t) = \lambda$, a constant. Thus, the problem of estimation would be an ordinary multiple linear regression. Of course, if the base hazard function is not a constant, then the regression would be more complicated. Cox (1972) used a maximum likelihood estimation procedure, that relies on numerical methods to solve the log-likelihood equations for the unknown parameters. These equations are independent of $h_0(t)$, so that they do not require any particular form of the base hazard. The accounting for censored observations is done through a method such as the Kaplan–Meier survival probability estimation process.

Key Points for Chapter 13

- Time-to-event data are usually characterized by the probability that the time will meet or exceed a particular time.
- The function describing the probabilities over time is often called the survival function or survival curve, $S(t)$.
- The survival function has several related functions that are unique to a specific survival curve.
- The most important related functions are the hazard function, $h(t)$, and its integral, the cumulative hazard function, $H(t)$.
- Empirical data can be used to estimate $S(t)$, $h(t)$, and $H(t)$.
- Often time-to-event data are censored; typically censoring happens when for some experimental unit, the event is never observed.
- The Kaplan–Meier method may be used to estimate $S(t)$ in the face of censored observations.
- There are several methods for comparing survival curves between treatment groups, including the Logrank test, the Peto and Peto test, and the Cox regression model.

Exercises and Questions

1. If $S(t) = e^{-\lambda t}$ then what is an expression for the average time-to-event?
2. What is an expression for the $100p^{th}$ percentile of an exponential time-to-event distribution with rate parameter λ? (Hint: Let $p = e^{-\lambda t}$ and solve for t.)
3. Suppose N = 10 time to event (TTE) observations were made, and contained in the file "20170619 Exercise 13.3 TTE.csv." This file has two variables, TTE and Censor, which indicates whether an observation was associated with an event (Censor = 1) or if it was a censored time (Censor = 0). Perform a Kaplan–Meier analysis using the Censor variable, and then the same analysis but with no censoring (e.g., make a new censoring variable that is 1 for all 10 observations). Is there a difference in the time-to-event probabilities?

14

Time Series Analysis and Stochastic Processes

General Ideas

A stochastic process is a sequence of random variables, ordered usually in time, but ordering could be along a spatial continuum. The variables can be continuous or discrete, and time (the ordering parameter) can be continuous or in discrete steps. We will discuss two particular types of stochastic processes:

1. Time series: Continuous variables observed over continuous time, but only recorded over a sequence of discrete points in time that are (more or less) uniformly spaced.
2. Markov chains: A particular class of discrete variables observed over a sequence of discrete steps. Most of the time, the variables will only have a finite number of possible values, called *states*.

The utility of these two kinds of stochastic processes is that by fitting models, the researcher can compute predictions, or forecasts, of the value of the sequence at future times (time series) or predict the likelihood that the process will be in any particular state in the future.

Time Series

Time series will be defined here as a sequence of continuously valued random variables. For this text, we will assume that the variables are all normally distributed. The interesting part is that the values in the sequence are assumed to have non-zero correlations with each other. That is, the value observed at time t will have some correlation with previous values. These correlations will be exploited to help identify the underlying relationship that the values have with each other over time. Since the sequence is generally a set of values of the same measurement, but at different times, the correlations will be referred to as autocorrelations.

Suppose we observe a variable at uniformly-spaced time intervals, for n intervals, and represent these values as

$$y_1, y_2, y_3, \ldots, y_t, y_{t+1}, \ldots, y_n$$

Mathematical statisticians would think of this as a realization of a sequence of random variables (Grimmet and Stirzaker, 2004). That is, suppose Y_1 was the first variable in the sequence. The value observed for Y_1 would be y_1. However, before Y_1 is observed, there are an infinite number of possible values it might have. Once Y_1 is observed, we think of it as being "realized." Suppose we further suspect that the value of y at time t is some unknown function not of time itself, but of the previous values of y, such as

$$y_t = \beta_0 + \beta_1 y_{t-1} + \beta_2 y_{t-2} + a_t$$

The β_i are (unknown) constants, and a_t is a random "noise" variable with a mean value of 0 and a standard deviation, σ, that does not change with time. Furthermore, the a_t are assumed to be independent of each other as well as identically distributed. We will almost always assume that the noise variable is normally distributed, but the fixed mean and standard deviation assumptions are the most critical. It is also possible that the value of y at the "current" time, t, is expressed as a function of noise values at previous times:

$$y_t = \alpha_0 + \alpha_1 a_{t-1} + \alpha_2 a_{t-2} + a_t$$

Again, a_t is a random "noise" variable, and the α_i are constants. Think of the values a_{t-1} and a_{t-2} as random variables whose values have (somehow) already been observed and recorded. The values of a_t and y_t have not yet been observed. We will restrict the discussion to those time series that are a linear function of previous values and noise values.

Both of the expressions described above are what we will call *second order*, since they only reach two steps back in time. The first expression is referred to as an autoregressive model (symbolized as AR); the second is referred to as a moving average (symbolized as MA) model (Cryer, 1986). It is also possible that the future values of the series are related to both the series past values and the past values of the random "noise":

$$y_t = \delta + \beta_1 y_{t-1} + \beta_2 y_{t-2} + \alpha_{t-1} a_{t-1} + \alpha_{t-2} a_{t-2} + a_t$$

Such a hybrid expression, or model, is called an *autoregressive moving average* (ARMA). The order of the models (how far back they reach in time) is designated with integers, for example, AR(2) is a second-order autoregressive model, MA(2) is a second-order moving average model, and ARMA(2,2) is an ARMA model of orders two and two.

Identifying Time Series Model Types and Orders

Box and Jenkins (Box, Jenkins, and Reinsel, 2008) discovered a methodology for identifying an appropriate model for time series. Their method involves computing two functions of the series data, called the autocorrelation function (*acf*) and partial autocorrelation function (*pacf*). First, some definitions are required. Recall the notion of expectation, or expected value. The covariance of two random variables, X and Y, is was defined earlier to be

$$Cov(X,Y) = E\left[(X - \mu_X)(Y - \mu_Y)\right] = \int_{-\infty}^{+\infty}\int_{-\infty}^{+\infty} (x - \mu_X)(y - \mu_Y)f(x,y)\,dx\,dy$$

The correlation between X and Y is a sort of standardized covariance:

$$Corr(X,Y) = \frac{Cov(X,Y)}{\sigma_X^2 \sigma_Y^2}$$

One could show that the correlation function is always a number between –1 and +1. In general, the correlation is positive if one variable increases when the other increases. It is negative if one variable decreases when the other increases. For time series, we are interested in how the values of the variable are related to values of the same variable, but observed at different times. So, for example, consider the values of a series at time t and at time $t–k$. The covariance would be

$$Cov(Y_t, Y_{t-k}) = E\left[(Y_t - \mu_{Y_t})(Y_{t-k} - \mu_{Y_{t-k}})\right]$$
$$= \int_{-\infty}^{+\infty}\int_{-\infty}^{+\infty} (y_t - \mu_{Y_t})(y_{t-k} - \mu_{Y_{t-k}})f(y_t, y_{t-k})\,dy_y\,dy_{t-k}$$

Since this covariance is using the same variable, just observed at different times, it is called the autocovariance. Accordingly, the autocorrelation is

$$Corr(Y_t, Y_{t-k}) = \frac{Cov(Y_t, Y_{t-k})}{\sigma_{Y_k}\sigma_{Y_{t-k}}}$$

In this text, we will only concern ourselves with time series that have a particularly convenient property; namely, that the autocorrelation only depends on the length of time between the time points, k, and not the specific time itself, t. Such time series will be referred to as *stationary*. This is not the strictest, most general definition of stationarity, but it is the one on which we will rely. Stationarity is a necessary condition for being able to identify the particular form and order of model to be fit to the data. The subscript k is called the *lag*. If $k = 0$, the autocorrelation is 1, and the covariance is the variance. For stationary time series, the covariance of lag k could be estimated as

$$\widehat{acov}(k) = \frac{\sum_{t=k+1}^{n} (y_t - \bar{y})(y_{t-k} - \bar{y})}{n}$$

This expression is only a function of the lag, k. That is, the assumption of stationarity is required for this estimator to be meaningful. Also, the average, $\bar{y}$, is taken over all the observations, y_t, $t = 1, n$. Recall that this estimator is only a reasonable estimator if the time series is stationary.

The autocorrelation function (*acf*) can be estimated as

$$r(k) = \widehat{acf}(k) = \frac{\widehat{acov}(k)}{\hat{\sigma}^2}$$

where

$$\hat{\sigma}^2 = \frac{\sum_{t=1}^{n} (y_t - \bar{y})^2}{n}$$

A function of the lag that is closely related to the *acf* is called the partial autocorrelation function (*pacf*), mentioned earlier. The *pacf* is a sort of conditional *acf*, and in general is the conditional expectation (Cryer, 1986):

$$\phi(k) = pacf(k) = acf\left(k \mid y_{t-1}, y_{t-2}, \ldots, y_{t-k+1}\right)$$

The estimated *pacf* values are obtained from the sample estimates of the *acf*, via a system of equations called the Yule–Walker equations (Box et al., 2008):

$$r(j) = \varphi_1 r(j-1) + \varphi_2 r(j-2) + \ldots + \phi(k)r(0), \ j = 1, \ k$$

That is, the coefficient for the zeroth value of the *acf* is the estimate of the *pacf* at lag k. Of course, the zeroth value of the *acf* is always one. The unknown coefficients, φ_i, are actually the coefficients for an autoregressive model. The important point is that the acf and pacf are functions of the lag for stationary time series. The estimates of these functions will be used to identify a model to fit to the data.

The Box–Jenkins Approach

Box and Jenkins discovered that different forms of ARMA times series had specific morphologies for their associated *acf* and *pacf* functions. Thus, if one can estimate the *acf* and *pacf* functions of a time series, one can then determine the appropriate time series model to fit to the data. Moving average models of order k (MA(k)) models have *acf* functions that "cut off" (i.e., drop to zero) after lag k, and *pacf* functions that "taper off" toward zero as the lag gets larger and larger. Autoregressive time series of order k (AR(k)) will have *acf* functions that taper off to zero as the lag gets large, and *pacf* functions that will drop to zero after lag k. An autoregressive, moving average model of orders k and l (ARMA(k,l)) will have a *pacf* function that drops to zero after lag k and an *acf* function that drops to zero after lag l.

To illustrate the *acf* and *pacf* of various time series, we simulated four different time series. The first is called "white noise"; it is a nonautocorrelated sequence of random variables all having exactly the same distribution, generally with mean zero and some standard deviation. If the variables are normally distributed, as in the case simulated here, the sequence is referred to as Gaussian white noise. In the case of white noise, we would expect the values of the *acf* and *pacf* to be patternless and relatively small. Figures 14.1 and 14.2

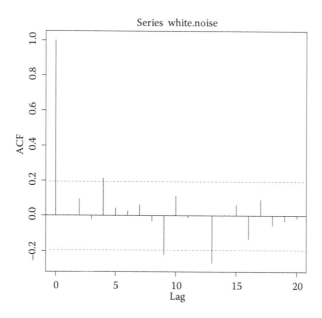

FIGURE 14.1
Acf of white noise.

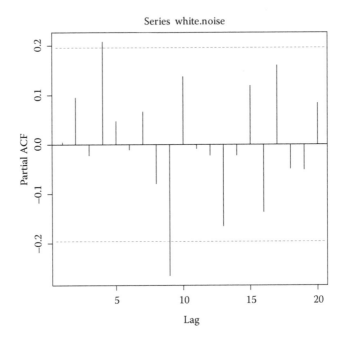

FIGURE 14.2
Pacf of white noise.

show the *acf* and *pacf* for the white noise sequence. The dashed horizontal lines are lower and upper limits of approximate 95-percent confidence intervals for the *acf* and *pacf*, under the hypothesis that the true values are zero.

The next series is purely autoregressive and of order 2, referred to as AR(2). Figures 14.3 and 14.4 show the acf and pacf, respectively.

The *acf* for the AR(2) series ought to oscillate and be "damped"; that is, gradually decrease in magnitude as lag increases. The *pacf* ought to be "large" for lags 1 and 2, and then become small for lags 3 and beyond. The lag 0 *acf* is always equal to 1.0. There is no lag 0 *pacf* value, inasmuch as *pacf* must be conditioned on previous lags, and nothing comes before lag 0. Technically, negative lags are possible in some cases, but we will not consider them here.

The *acf* and *pacf* for a pure moving average process of order 2, MA(2), are shown in Figures 14.5 and 14.6, respectively.

It is not necessarily clear that an *acf* or *pacf* "cuts off" at a particular lag, or is "damped."

Finally, Figures 14.7 and 14.8 show the *acf* and *pacf* for a mixed autoregressive, moving average model or order 2,2, or ARMA(2,2). In theory both the *acf* and *pacf* should drop to 0 (i.e., fall within the 95-percent confidence limits) after lag 2.

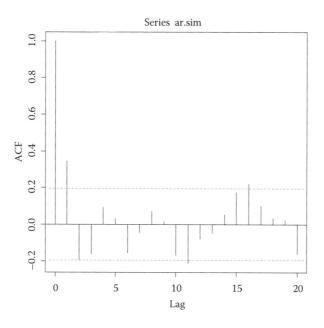

FIGURE 14.3
Acf of AR(2).

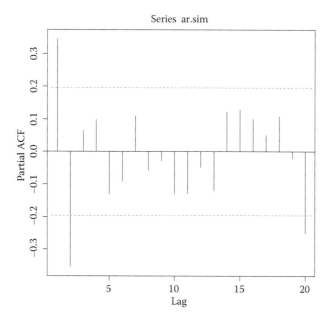

FIGURE 14.4
Pacf of AR(2).

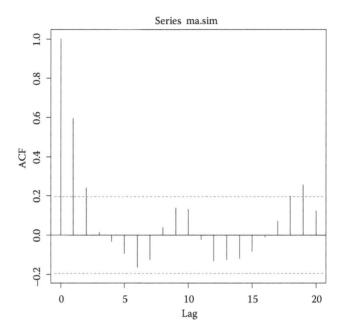

FIGURE 14.5
Acf for MA(2).

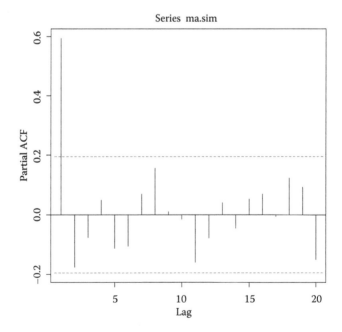

FIGURE 14.6
Pacf for MA(2).

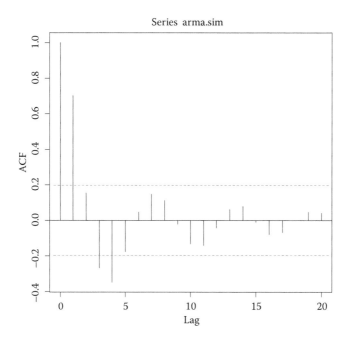

FIGURE 14.7
Acf for ARMA(2,2).

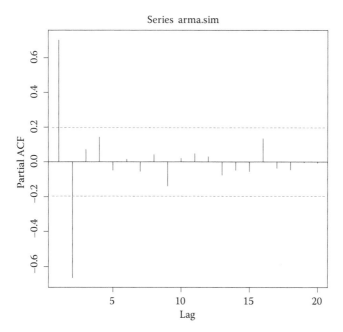

FIGURE 14.8
Pacf for ARMA(2,2).

Nonstationarity and Differencing

One of the primary assumptions required for the Box–Jenkins model identification process is stationarity. There is however a class of nonstationary time series where stationarity can be induced via a simple transformation called differencing. If Y_t is a time series, then define the difference of order 1 to be

$$D_t = Y_t - Y_{t-1}$$

A difference of order 2 would be a difference of differences:

$$D_t^{(2)} = D_t - D_{t-1}$$

In general, an order k difference would be

$$D_t^{(k)} = D_t^{(k-1)} - D_{t-1}^{(k-1)}$$

There are situations where such differenced series are stationary although the original series are not. Consider, for example, the following series:

$$y_t = y_{t-1} + \beta_2 y_{t-2} + a_t$$

This series is in fact nonstationary. However, the series

$$d_t = y_t - y_{t-1} = \beta_2 y_{t-2} + a_t$$

may in fact be stationary, depending on the value of β_2. An ARMA model that is based on k^{th} order differences is called an ARIMA model, where the "I" stands for "integrated." The "integration" is how the predicted values of the original series are obtained from the model fit to the differences.

Examples with R Code: Time Series

Time Series

The hairy-faced hover wasp (*Liostenogaster flavolineata*), is a facultatively social species found in tropical Southeast Asia. In this species, colonies consist of a single dominant egg-laying female, and multiple nonlaying females who are the daughters or younger sisters of the dominant egg-layer. Newly eclosed (i.e., emerged from the pupal stage) females can either choose to stay in their natal nest as a nonbreeding helper, with the possibility of inheriting the dominant position in the future, or choose to disperse. The costs and benefits of staying versus dispersing are likely affected by a number of factors, including the expected size of the colony around the time when the female would be likely to inherit. If there are predictable seasonal patterns in brood production, then females could potentially use seasonal or climatic cues to inform the decision of whether to stay or disperse.

Cronin et al. (2011) investigated seasonal patterns in brood production in a population of hairy-faced hover wasps in Peninsular Malaysia. In this region, rainfall exhibits two annual peaks: One during April, and one during October. Once every ten days between April and October, the authors censused the number of adults and brood within a variable number of colonies. Here, we provide an example based on this study to illustrate the techniques involved in analyzing time-series data. While Cronin et al. were interested in the effects of season, rainfall, and temperature on the numbers of new wasps (larvae and pupae), we will simply examine the possibility of observing new larval counts over time. We will treat the count of larvae, averaged over multiple colonies at each time point, as a continuous variable. To identify an appropriate model for the time series, we will use the R functions *acf*() and *pacf*(). Figures 14.9 and 14.10 show the *acf*() and *pacf*() function output, respectively.

The first three non-zero "lags" in the *acf* have values outside the dashed lines, and the first two "lags" in the *pacf* are outside the dashed lines. Based

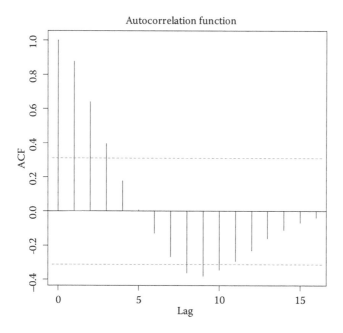

FIGURE 14.9
Acf output: Larval counts.

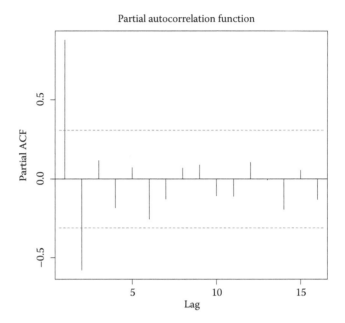

FIGURE 14.10
Pacf output: Larval counts.

on those results, we might guess that an appropriate model would be ARMA(2,3), that is, a two-lag autoregresssive part and a three-lag moving average part.

The model was fit with two different R functions, *arima*(), which is in base R, and *Arima*(), which is a function in the package "forecast." The code and associated R console output are shown in Figure 14.11.

The console output for *arima*() and *Arima*() are nearly identical. The model parameter estimates, or coefficients, are shown, together with their standard errors. In this case, the estimated model is

$$y_t = 8.9416 + 0.8763y_{t-1} - 0.2325y_{t-2} + 0.5968a_{t-1}$$
$$+ 0.2521a_{t-2} + 0.2508a_{t-3} + a_t$$

where y_t is the count at time t and a_t is the random fluctuation or noise at time t.

Arima() also produces an object having the fitted values as an attribute (arma.model2$fitted), which can then be compared to the observed values over time. Figure 14.12 shows the plot.

In the *arima*() or *Arima*() functions, the model order is specified by three numbers: The autoregressive order, the differencing order, and the moving average order. The orders are specified in that way: order=c(a,d,m). If any

```
setwd("C:\\Users\\SMFSBE\\Statistical Data & Programs")
df1 <-read.csv("20170716 Example 14.1 Wasp Larvae Counts Time Series.csv")
#
#library(forecast)
#
#Variables:
#
# time.index
# Larvae
#

attach(df1)

acf(Larvae,main="Autocorrelation Function")
dev.new()
pacf(Larvae,main="Partial Autocorrelation Function")
dev.new()
arima(Larvae,order=c(2,0,3))
arma.model2 <-Arima(y=Larvae,order=c(2,0,3))
plot(x=time.index,y=Larvae,main="Larval Counts over Time
(weeks)",xlab="Week",pch=1,col=1)
points(x=time.index,y=arma.model2$fitted,pch=17,col=2)
legend(x=5,y=5,legend=c("Observed","Fitted"),pch=c(1,17),col=c(1,2))


OUTPUT:

Call:
arima(x = Larvae, order = c(2, 0, 3))

Coefficients:
         ar1         ar2      ma1      ma2      ma3     intercept
      0.8763     -0.2325   0.5968   0.2521   0.2508       8.9416
s.e.  0.6747      0.4362   0.6376   0.6682   0.3730       1.0834

sigma^2 estimated as 1.486:  log likelihood = -65.9,  aic = 145.81
```

FIGURE 14.11
Arima modeling: Wasp larvae counts.

number is zero, it indicates that this facet of the ARIMA paradigm is not part of the model. Thus, order=c(2,0,3) means fit a model with two autoregressive terms, zero differencing, and three moving average terms.

Markov Chains

A Markov chain is a sequence of discrete random variables, $X_1, X_2, \ldots$, changing in discrete time steps, that has a special property:

$$Pr\{X_k = x_k | X_0 = x_0, \; X_1 = x_1 \ldots X_{k-1} = x_{k-1}\} = Pr\{X_k = x_k | X_{k-1} = x_{k-1}\}$$

In other words, the probability that the current value in the sequence has any particular value only depends on the immediately preceding value, and not on the entire history of the sequence. This property is called the *Markov property* (Grimmett and Stirzaker, 2004).

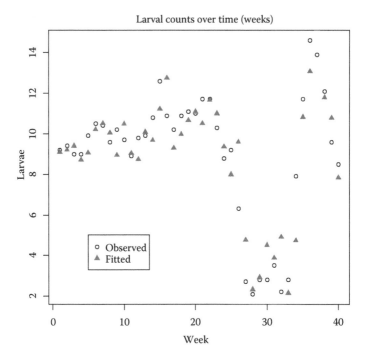

FIGURE 14.12
Larval count data: Observed and fitted values over time.

For the sake of simplifying notation, we will refer to the values or "states" of the X_k as 1, 2, 3, ..., that is, $X_k = 2$, although there is not necessarily any order to the states. For example, a traffic light can have the "states" of red, yellow, or green. We could arbitrarily assign to the "red" state the value one, "yellow" could be two, and "green" could be three. For further simplification, we will use the notation for the conditional probabilities:

$$Pr\{X_k = j \mid X_{k-1} = i\} = p_{ij}$$

For processes having the Markov property, p_{ij} is the conditional probability that the next state will be j given that the current state is i. These conditional probabilities are called transition probabilities. Most of the chains we will discuss have only a finite number of possible states, as in the case of traffic lights. Thus, the one-step transition probabilities for an n-state chain can be represented by a (finite-dimensional) $n \times n$ matrix:

$$P = \begin{bmatrix} p_{11} & p_{12} & \cdots & p_{1n} \\ \vdots & \ddots & \ddots & \vdots \\ p_{n1} & p_{n2} & \cdots & p_{nn} \end{bmatrix}$$

As we have defined the transition probability matrix, each row represents a conditional distribution for the states $j = 1, n$, that is,

$$\sum_{j=1}^{n} p_{ij} = 1$$

Keep in mind that it is thoroughly possible that there are some states such that:

(1) $p_{ij} = 1$
(2) $p_{ij} = 0$
(3) $p_{ij} = 0$ for $j \neq i$ and $p_{ii} = 1$

The condition in (3) is referred to as "absorbing," so that state i is called an "absorbing state." That is, once the chain reaches state i it can never transition to any other state.

There are several interesting results and consequences related to finite-state Markov chains. It all stems from the Markov property. The matrix P is called the one-step transition probability matrix. The transition probabilities for two, three, four, and so on steps ahead can be calculated as a power of matrix P. If $p_{ij}(1)$ represents the transition probability from state i to state j in one time step, and $P(k)$ is the matrix of k-step transition probabilities, then

$$P(k) = P^k$$

In many cases, as k gets very large, there is a convergence so that each row is identical, and the row represents the long-run probability of finding the "system" in each of the states. Such a long run probability distribution is called the steady-state distribution, and can be calculated. From this steady state distribution, the expected state can also be calculated.

Extensions of Markov Chains

There are many extensions to the ideas of Markov chains. We will only mention two that have gained some popularity: Higher-order chains and hidden Markov models (HMMs). As stated earlier, the Markov property for a stochastic process states that the probability of the chain being in a certain state at time k depends entirely on the state it was in at the immediately preceding time, k–1. A higher-order chain is one in which that dependence is extended to k–1, k–2, …, k–m. So, for a second-order chain,

$$Pr\left\{ X_k = x_k \middle| X_0 = x_0, \ X_1 = x_1 \dots X_{k-1} = x_{k-1} \right\}$$
$$= Pr\left\{ X_k = x_k \middle| X_{k-1} = x_{k-1}, \ X_{k-2} = x_{k-2} \right\}.$$

When such higher-order dependencies exist, the properties that make Markov chains so useful (e.g., finding stationary distributions) disappear. Thus, computing with higher-order processes becomes far more difficult. One possible way of finding limiting distributions of a higher-order chain's states is via simulation. This, however, is beyond the scope of this text.

Another extension is the notion of hidden Markov models. A hidden Markov model (HMM) is one whose states are not directly observable, but they depend on another chain whose states are observable. Suppose, for example, the change from season to season in the population size of some kind of beetle cannot be assessed, but the change in the population size of a bird that feeds on those beetles can. If the state is population size, and the population size of the birds changes in a Markov fashion, then perhaps the change in the beetle population size can be assessed indirectly.

While these extensions are beyond the scope of this book, the interested reader can find a straightforward tutorial in Fosler-Lussier (1998).

Examples with R Code: Markov Chains

Many behaviors occur in predictable sequences in which the probability of occurrence of one element depends on previous elements, and as such, can be modeled using Markov chains. This is especially true of sexually selected displays, such as songs or courtship dances, which often consist of multiple discrete elements arranged in non-random fashion. While the description of such displays is by no means the only application of Markov chains within the field of behavioral ecology, it is a very common one. The following example uses the courtship displays of wolf spiders to illustrate the utility of Markov chains for describing animal behavior, and is based on a study by Chiarle and Isaia (2013).

Pardosa proxima and *Pardosa vlijmi* are two sister species of wolf spider that are essentially indistinguishable based on morphology, but nonetheless show markedly distinct courtship displays. Both species begin their display with a "starting" unit that is only performed once, although the actual form of the starting unit differs between the two species. The starting unit is followed by the main unit, which is in turn composed of two phases that are repeated in a cyclic fashion. In *P. proxima* the two phases are called "stepping" and "jerking," while in *P. vlijmi* they are called "springing" and "swinging." Finally, the main unit of the display is followed by the ending unit, which is only performed once. Chiarle and Isaia (2013) collected videos of male spiders from both species performing their courtship displays, and calculated the transition probabilities between each stage in the display.

In the example, we will assume four possible states:

S: "Starting"
M1: One of the two "main unit" states ("stepping," say, for *P. proxima*, or "springing," for *P. vlijimi*)
M2: The other "main unit" state ("jerking" for *P. proxima*, or "swinging" for *P. vlijimi*)
E: "Ending"

Table 14.1 shows the matrix of transition probabilities we will use for the example (only vaguely similar to those observed by Chiarle and Isaia, and only intended to illustrate the Markov chain computations).

We will focus on computing a stationary distribution for the states. Figure 14.13 shows the R code and output. The computation of the stationary distribution involves some iteration on choosing a power k that yields identical values in each row of the resulting matrix P^k.

Thus, the estimated "long-run" probabilities of observing a spider in any of the four states are

$Pr\{S\} \approx 0.804$ percent
$Pr\{M1\} \approx 0.965$ percent
$Pr\{M2\} \approx 0.965$ percent
$Pr\{E\} \approx 97.267$ percent

Of course, computing these is entirely dependent on the process having the Markov property, and the one-step transition probability matrix.

Knowing the stationary distribution is useful if knowing that the probability of finding the chain in a particular state at any time in the future is helpful. Another useful fact is that the mean recurrence time, or time until the chain returns to a state, is the reciprocal of the stationary probability for the state. Thus, the average number of time steps to return to state S, for

TABLE 14.1

One-Step Transition Matrix

	S	M1	M2	E	Sum
S	0.879	0.060	0.060	0.001	1.000
M1	0.000	0.110	0.840	0.050	1.000
M2	0.000	0.840	0.110	0.050	1.000
E	0.001	0.000	0.000	0.999	1.000

```
setwd("C:\\Users\\SMFSBE\\Statistical Data & Programs")
df1 <-read.csv("20170717 Example 14.2 Transition Probability Matrix.csv")
#
#
# VARIABLES:
# S
# M1
# M2
# E
#
attach(df1)
Amat <-as.matrix(df1)
Smat <-eigen(Amat)
Smat$values
Smat$vectors
Lamda <-diag(Smat$values)
#
# This will compute the stationary distribution (if it exists)
# of a finite, irreducible Markov Chain
# if the input matrix is a one-step transition matrix
# for that chain
#
n <-455
Lamdan <-diag(Smat$values**n)
Eig <-as.matrix(Smat$vectors)
EigInv <-solve(Eig)
Arecover <-Eig%*%Lamda%*%EigInv
An <-Eig%*%Lamdan%*%EigInv
Lamda
Eig
Amat
Arecover
n
An
#
# rows of An will be the stationary distribution if n is large enough
```

FIGURE 14.13
R code for computing stationary distribution of Markov chains.

example is $1/0.00804 \approx 124$ steps. Table 14.2 shows the stationary probabilities for the four states in the example, together with their mean recurrence times.

On the average, if the chain is observed to be in state M1, for example, then it will be in state M1 again in 104 time steps.

It may be of interest to compare stationary distributions for two chains. Suppose that the empirically observed transition matrix for *P. vlijimi* is as shown in Table 14.3, and the transition matrix for *P. proxima* in Table 14.4. Again, these are not numbers actually recorded in the work by Chiarle and

TABLE 14.2

Stationary Probabilities and Associated Mean Recurrence Times

State	Stationary Probability	Mean Recurrence Time
S	0.00804	124
M1	0.00965	104
M2	0.00965	104
E	0.97267	1

TABLE 14.3

State Transition Frequencies: *P. vlijimi*
One-Step Transition Probabilities *P. vlijim*

	S	M1	M2	E	Sum
S	0.010	0.087	0.087	0.816	1.000
M1	0.020	0.029	0.880	0.071	1.000
M2	0.040	0.850	0.039	0.071	1.000
E	0.001	0.000	0.000	0.999	1.000

TABLE 14.4

State Transition Frequencies: *P. proxima*
One-Step Transition Probabilities *P. proxima*

	S	M1	M2	E	Sum
S	0.014	0.085	0.085	0.816	1.000
M1	0.040	0.850	0.039	0.071	1.000
M2	0.020	0.029	0.880	0.071	1.000
E	0.002	0.000	0.000	0.998	1.000

Isiai; these are just numbers used for illustration. Suppose further that there were $n = 200$ observations of the states for each species at randomly selected observation times. The numbers of observations in each state for each species are given in Table 14.5.

There are two questions to ask:

1. What is the estimated stationary distribution for each species?
2. Does it appear that these two distributions differ?

To answer the first question, use the method implemented in the code shown in Figure 14.13. To answer the second, use the methods described in Chapter 7 (chi-squared test for independence). These tasks are left as exercises for the reader.

Figure 14.14 is an illustration of the hairy-faced hover wasp.

TABLE 14.5

Counts of State Observations for Each Species

State	P. vlijimi	P. proxima
S	2	3
M1	1	2
M2	3	5
E	194	190

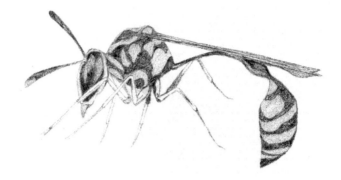

FIGURE 14.14
Hairy-faced hover wasp (*Liostenogaster flavolineata*).

Theoretical Aspects

Time Series

The relationship between a time series and its associated difference series may not be perfectly obvious. Consider the series:

$$y_t = y_{t-1} + \beta_1 y_{t-1} + a_t$$

And its associated first-difference series:

$$d_t = y_t - y_{t-1} = \beta_1 y_{t-1} + a_t$$

Suppose the initial observed value of the series is y_0. Then

$$y_1 = y_0 + \beta_1 y_0 + a_1$$

$$y_2 = y_1 + \beta_1 y_1 + a_2$$

$$d_2 = y_2 - y_1 = y_1 + \beta_1 y_1 - y_0 - \beta_1 y_0$$
$$+ a_2 - a_1 = y_1 - y_0 + \beta_1 (y_1 - y_0) + a_2 - a_1$$

or

$$d_2 = d_1 + \beta_1 d_1 + b_2$$

with $b_2 = a_2 - a_1$. Due to the assumptions of independence and identical distributions for all the a_t, b_2 (and by induction, all b_t) also have mean zero with a constant standard deviation. By induction:

$$d_t = d_{t-1} + \beta_1 d_{t-1} + b_t$$

The point is that the form of the model and its parameters estimated using the differenced series are in fact those of the original series.

Differenced series are referred to as "integrated" (Box et al., 2008). Thus

$$d_2 + d_1 = y_2 - y_1 + y_1 - y_0 = y_2 - y_0$$

More generally:

$$d_t + d_{t-1} = y_t - y_{t-1} + y_{t-1} - y_{t-2} = y_t - y_{t-2}$$

Summing, or "integrating" up to time t gives:

$$\sum_{k=1}^{t} d_k = \sum_{k=1}^{t} (y_k - y_{k-1}) = y_t - y_0$$

Of course, if the initial condition is that $y_0 = 0$, the integration of the differences yields the original value of the series at time t. The value of y_0 is simply the constant of integration.

Markov Chains

We noted that the k-step transition matrix for a Markov chain is simply the kth power of the one-step transition matrix:

$$P(k) = P^k$$

This fact leads to a convenient "updating" formula. That is, if $P(k)$ is known, then $P(k+1)$ can be computed as

$$P(k+1) = PP^k$$

In general:

$$P(k+m) = P^k P^m$$

This relationship is called the *Chapman–Kolmogorov equation* (Grimmett and Stirzaker, 2004).

To obtain the stationary distribution, first obtain the eigenvalues, and the associated matrix of eigenvectors of the transition matrix. The stationary distribution is the limit:

$$\Pi = \lim_{k \to \infty} B \Lambda^k B$$

Where Λ is the diagonal matrix of eigenvalues, and B is the matrix of eigenvectors of the transition matrix. Each row of matrix Π will be identical, and will be the stationary distribution for the chain, if such a distribution exists.

Key Points for Chapter 14

- A stochastic process is a (potentially infinite) sequence of random variables.
- The random variables in a stochastic process may be discrete or continuous.
- A time series is a stochastic process that is a sequence of (usually) continuous variables observed at uniformly spaced points in time.
- Time series are characterized by their autocorrelation and partial autocorrelation functions, which quantify the dependence of the series value at time t on previous values.
- A time series is stationary if the autocorrelation and partial autocorrelation functions only depend on the difference in time between two values in the sequence (how far apart two values are in time), and not the actual time itself.
- Stationary time series models can be identified by patterns in the autocorrelation and partial autocorrelation functions, which both can be estimated using an observed sequence of data.
- In many cases, stationarity can be induced by using differences rather than the original series to identify and fit models.
- A fitted time series model can be used to forecast future values; the stationary distribution of a Markov chain can be used to predict the likelihood of each state, and the expected number of steps to revisit any state.
- A Markov chain is a sequence of discrete variables, also observed at discrete steps, often time steps. To be a Markov chain it must have the Markov property.

- The Markov property relates to conditional probability distributions; the probability that the next variable in the sequence has a particular value is only dependent on its current value, and not on the entire history of the sequence.
- Markov chains can be characterized by a matrix of probabilities that the variable value at one step will then be a different value in the next step; this matrix is called the *transition matrix*.
- The transition matrix can be used to compute a "stationary" distribution, or the long-run average probabilities that the chain would contain any particular value, or state.

Exercises and Questions

1. Suppose the transition matrix for the wolf spider example was

	S	M1	M2	E	Sum
S	0.100	0.330	0.330	0.200	1.000
M	0.050	0.030	0.060	0.860	1.000
M2	0.100	0.200	0.300	0.400	1.000
E	0.010	0.050	0.050	0.890	1.000

Compute the stationary distribution and mean recurrence times for the four states.

2. Use the *forecast()* function in package forecast, together with an *Arima()* object, to forecast the time series model fit to the wasp larvae count data for h = 3 periods beyond the end of the data. In addition, try fitting time series models of order (2,0,2), (1,0,1) and (2,0,0), and compare the forecasts (informally).

15

Study Design and Sample Size Considerations

Study design can be partitioned into two basic aspects, the procedural aspects and the analytic aspects. By procedural we mean those things that must be done in the course of the study, such as capturing individuals, marking them, recording them, or other physical processes required to obtain observations. By analytical aspects, we mean those strategic elements that determine how the data will be organized for later analysis, such as deciding which groups of individuals to sample, the order in which treatments will be applied to experimental units, or the factors that will be varied in the course of the experiment. The focus here will be entirely on the analytical aspects.

Degrees of Freedom: The Accounting of Experimental Design

Degrees of Freedom are numbers that indicate the number of terms in a sum that can vary freely and still allow the sum to have a particular value. Thus, the total degrees of freedom in an analysis will be the entire sample size (total number of observations) minus one. The notion is that if the sum of all observations was known to be a particular value, then the first $n - 1$ values could be any number, but the last term in the sum would be completely determined by the values of the previous terms.

In an experiment in which some number of treatments are applied, there may be effects other than the treatments that could affect the response values, but are not of interest per se. Rather, they interfere with the experimenter's ability to determine if the treatments have differential effects on the response. Suppose, for example, that an experimenter wants to determine if animals respond differently to three different stimuli. The problem is that even with the same stimulus, two individuals might respond differently. Suppose further that the plan is to expose each subject to each of the three stimuli. For now assume that the order of exposure has no effect, and there is no carry-over; that is, the individuals have no memory of previous stimuli. Nevertheless, two individuals could have different "average" responses in general, but the effects of the stimuli could still be consistent (e.g., stimulus A increases the response above the individual's average, stimulus B decreases the response below the individual's average, and stimulus C has no effect). If there are $n = 20$ individuals, there are a total of 3*20 = 60 observations. Thus, the total degrees of freedom in this experiment are 20 – 1 = 19. Based on the assumptions of linearity in terms of the effects, the degrees of freedom can be partitioned into the following categories:

Individuals

Treatments

Noise, or Error

Total

The degrees of freedom are based on the average differences between each value or level of the effect and the overall average. Thus, there are 20 Individuals, so there would be 20 Individual averages, and 20 differences from the overall average. Thus, there are 20 – 1 = 19 degrees of freedom for Individuals. Since there are 3 Treatments, there are 3 – 1 = 2 degrees of freedom for Treatments. The Noise or Error degrees of freedom are obtained by subtraction of the sum of degrees of freedom for all the effects from the total degrees of freedom. Thus,

Error d.f. = Total d.f. – Individuals d.f. – Treatments d.f. = 59 – 19 – 2 = 38.

Latin Squares and Partial Latin Squares: Useful Design Tools

Suppose there are several treatments to be applied to each individual subject in a study. Furthermore, suppose that the order in which those treatments are applied might have some effect on the response. If there are three treatments to be applied, then there are 3! = 6 possible orderings of those treatments. If there are four treatments, then there are 4! = 24 possible orderings; five treatments would yield 5! = 120 orderings. It is likely that doing five treatments in all possible orderings is, to say the least, not practical. There is a solution that would allow the analysis to account for order effects as well as treatment effects, and individual effects, all in one ANOVA, without having to perform the experiment with an impractical number of subjects. Suppose there are four treatments. Consider the array of treatments, labeled A, B, C, and D, in Table 15.1. This table is referred to as a 4 × 4 Latin Square (Montgomery, 2001), since the treatments are symbolized with Latin letters, and they are arranged in a square of four rows and four columns. In each row and column, each treatment appears exactly once. The rows represent groups of subjects, assigned to have the treatments in the order specified by the row. Group I subjects would have

TABLE 15.1

A 4 × 4 Latin Square

		Order		
Group	1st	2nd	3rd	4th
I	A	B	C	D
II	B	C	D	A
III	C	D	A	B
IV	D	A	B	C

the treatments applied in the order A, B, C, and then D. Group II subjects would have the order B, C, D, and then A. If there are equal numbers of subjects in each group, then each treatment will be applied in each order an equal number of times.

Suppose there are n subjects in each group. The degrees of freedom for treatments is $4 - 1 = 3$. The degrees of freedom for Group is $4 - 1 = 3$. The total degrees of freedom would be $4*4*n - 1$ (four treatments by four groups by n subjects per group). The degrees of freedom for subjects is the number of groups times one less than the number of subjects per group, namely $4*(n - 1)$. The subjects are said to be "nested" inside Group. If Subject represented the individual subject identification code or name, and Group was the variable describing the group number, then in functions *lm()*, *glm()*, *aov()*, or *anova()*, the nesting can be represented as Subject%in%Group.

Suppose there are four treatments, but it is only possible to apply three treatments to any individual. A "rectangle" can be formed by simply eliminating the last column. The computation of degrees of freedom is unchanged, except that the total degrees of freedom would now be $4*3*n - 1$, since the total number of observations is now $4*3*n$. Such a rectangular array is called a Youden Square (Montgomery, 2001).

Such designs (Latin Squares, Youden Squares) and others are referred to as error control designs, since things like order of treatment are "nuisance" factors which must be accounted for, but are not usually of particular interest. These and other such designs, as well as formulae for degrees of freedom, may be found in Cochran and Cox (1957).

Power for ANOVA

The degrees of freedom for Error (dfe) are mainly responsible for the power of the test. That is, the larger the number of dfe the smaller the differences between effect levels that can be "detected." By detection, we mean that a difference will be identified as statistically significant with some stated probability. That probability is called *power*.

For ANOVA, the initial test for significance of any term in the linear model is based on the F ratio. The critical value comes from an F distribution with numerator degrees of freedom equal to those of the term (often called factor, or effect) of interest, and the denominator degrees of freedom are the dfe. The critical value is chosen so that if in fact the term has no effect on the response, there would be only about a $100\alpha\%$ chance of getting a sample F ratio exceeding this critical value. So, if there are three treatments, and all three are applied to each of twenty subjects, then there are $3 - 1 = 2$ degrees of freedom for treatments, and $60 - 1 - 2 - 19 = 38$ degrees of freedom for error (dfe). Thus, using the R function $qf()$, the critical value at the $\alpha = 0.05$ (Type I risk) level is

$$F(0.05,\ df1 = 2,\ df2 = 38) = qf(p = 0.05,\ df1 = 2,\ df2 = 38,\ lower.trail = FALSE)$$
$$= qf(p = 0.95,\ df1 = 2,\ df2 = 38,\ lower.trail = TRUE) = 3.244818$$

So, F(0.05, df1 = 2, df2 = 38) = 3.244818 is the 95th percentile of an F distribution with numerator degrees of freedom = 2 and denominator degrees of freedom = 3.244818.

In the ANOVA for this situation, the null hypothesis is that the effect of treatment is 0, that is, the response will not change (on the average) regardless of which treatment is applied. If the null hypothesis is correct, then there is about a 5-percent chance that a sample F statistic will exceed the value of 3.244818. But what if the null is incorrect? Perhaps one or more of the treatments would result in a slightly different average response compared to the others? Well, if the difference is very small, the probability of exceeding 3.244818 is not too much greater than 5 percent. Conversely, if the differences between at least two of the three treatments in terms of average response is actually large, then there would be a substantially higher probability of the sample F ratio exceeding 3.244818. Can this probability be computed? Yes, but only given knowledge of the within-subject standard deviation for the response, and assuming that this standard deviation is not different from one treatment to the next. The key is the ability to compute an extra parameter of the F distribution, known as *noncentrality* (Desu and Raghavarao, 1990). The noncentrality parameter is

$$\frac{\tau^2}{\sigma^2}\left(\frac{n}{df+1}\right)$$

τ represents the size of the effect (average difference in response).
σ is the "guess" at the within-subject standard deviation of the response.
n is the total sample size.
df is the number of degrees of freedom of the effect of interest (e.g., treatments).

Of course, even if a value of σ is known, the value of τ is not. Suppose

$$F\left(df1, df2, ncp = \frac{\tau^2}{\sigma^2}\left(\frac{n}{df1+1}\right)\right)$$ represents a random variable having an F

distribution with those associated parameters. A curve of probabilities can be constructed by computing:

$$Pr\left\{F\left(df1, df2, ncp = \frac{\tau^2}{\sigma^2}\left(\frac{n}{df1+1}\right)\right) > F(0.05, df1 = 2, df2 = 38\right\}$$

for varying values of $\tau \geq 0$. Such a curve is called a power curve.

Sometimes it is easier to specify the ratio $\dfrac{\tau}{\sigma}$ rather than just τ. The ratio is the number of standard deviations different from the overall average. Figure 15.1 shows R code for computing a power curve as a function of the ratio $\dfrac{\tau}{\sigma}$. Figure 15.2 shows the curve.

When the noncentrality parameter is 0, the F distribution is referred to as a "central" F. In this particular case, there is about 95-percent power to reject H_0 (the null hypothesis of no effect) if the effect has a magnitude of about 0.91 standard deviations.

In summary, the degrees of freedom for error (dfe) is the single most important number in determining power. The dfe are computed by subtracting the degrees of freedom for all the other terms in the ANOVA model from the total degrees of freedom, which in turn is the total sample size minus one. The power is computed using the non-central F distribution, where the noncentrality parameter is in part a function of how big an effect might exist.

```
setwd("C:\\Users\\SMFSBE\\Statistical Data & Programs")
#
#
#
tausigma <-seq(from=0.0,to=4.00,by=0.10)
nsubjects <-20
ntreatments <-3
nrep <-nsubjects * ntreatments
dfnum <-ntreatments -1
dfe <-nrep -1 -dfnum - (nsubjects -1)
noncent <-(tausigma**2)*nrep/(dfnum+1)# this is the set of non-centrality parameters
Fcrit <-qf(p=0.05,df1=dfnum,df2=dfe,ncp=0.0,lower.tail=FALSE)
Power <-pf(q=Fcrit,df1=dfnum,df2=dfe,ncp=noncent,lower.tail=FALSE)
plot(tausigma,Power,type="b",main="Power Curve for ANOVA",xlab="effect size, in no. of
standard deviations",ylab="Power = Pr{Reject H0}")
abline(h=0.05)
abline(h=0.95)
abline(v=0.91)
text(x=1.90,y=0.075,"Type I Risk = 5%")
text(x=1.8,y=0.975,"Power = 95%")
```

FIGURE 15.1
R code for computing power curve for ANOVA.

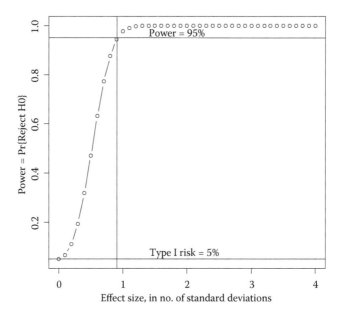

FIGURE 15.2
Power curve for ANOVA.

Sample Size and Confidence Intervals

With $n = 0$ observations, we can always state we have 100-percent confidence that anything we are trying to estimate lies somewhere in the interval $\pm\infty$. In fact, for most things to be estimated, we would be 100-percent confident that the parameter would fall somewhere between 0 and $+\infty$. In either case, neither of these confidence statements are likely to be sufficient, or helpful. In order to narrow the feasible range to something more useful, two things must change:

1. Give up the ability to say 100-percent confident.
2. Increase n.

The question is how precisely can we estimate a feasible range for the parameter of interest. For example, a confidence interval for the mean, or expected value, of a random variable, μ, is

$$\bar{x} \pm t_{\alpha,n-1} \frac{s}{\sqrt{n}}$$

$t_{\alpha,n-1}$ is the $100(1-\alpha/2)$%ile of a Student's t distribution with $n - 1$ degrees of freedom.

The "half-width" of the confidence interval is

$$t_{\alpha,n-1} \frac{s}{\sqrt{n}}$$

If we had a guess at the value of the standard deviation, call it $\hat{\sigma}$, and we wanted to know the value of the mean within an error range of $\pm\delta$, then we could solve for the required sample size:

$$n = \left(\frac{t\hat{\sigma}}{\delta} \right)^2$$

Since n must be an integer, we would generally truncate the result and add 1:

$$n = trunc\left(\frac{t\hat{\sigma}}{\delta} \right)^2 + 1$$

Furthermore, since t is a function of n, the sample size must be solved for by iteration.

As an alternative, Stein's two-stage approach (Desu and Raghavarao, 1990) could be employed. The idea is to first gather a small number of observations, say n_0. Then compute the standard deviation of the observations. Call this standard deviation s_0. Then, given the desired half-width δ, compute the required number of additional observations, n_1:

$$n_1 = trunc\left(\frac{t_{\alpha,n_0-1}s_0}{\delta} \right)^2 + 1$$

The total sample size would be $n_0 + n_1$. The value of t_{σ,n_0-1} is the $100(1-\alpha/2)\%$ile of a Student's t distribution with $n_0 - 1$ degrees of freedom.

Confidence Intervals for Proportions

The notion of sample size and confidence interval width applies to proportions as well as means. That is, the range of the lower and upper confidence limits for a proportion will get smaller as the sample size gets larger. The problem with proportions is that the confidence interval width depends on the proportion itself. Confidence interval widths for means do not depend

```
setwd("C:\\Users\\SMFSBE\\Statistical Data & Programs")
rm(X,N,alpha)
df1 <-read.csv("20170721 Example 15.2 Clopper Pearson Confidence Intervalscsv")
#
# X = no. of successes out of N
# N = sample size
# alpha = confidence coefficient (confidence = 1-alpha)
#
# ref:
# Clopper, C.J., Pearson, E.S., (1934) The use of confidence or
#                        fiducial limits illustrated in the case of the binomial
#                        Biometrika, 26, pp. 404-413
#

pL <-c()
pU <-c()
p <-c()
attach(df1)
alpha = 0.05
p <-X / N
nsets <-nrow(df1)

for (i in1:nsets){
 if (X[i] <= 0) {
#               note that the case X[i] < 0 is an absurdity
  pL[i] <-0
  }
  else {
  pL[i] <-(2*X[i])*qf(alpha/2,2*X[i],2*(N[i] -X[i] + 1))/(2*(N[i]-X[i]+1) +
(2*X[i])*qf(alpha/2,2*X[i],2*(N[i]-X[i] + 1)))
  }

 if (X[i] >= N[i]) {
#               note that the case X[i] > N[i] is an absurdity
  pU[i] <-1
  }
  else {
  pU[i] <-(2*(X[i]+1))*qf(1-alpha/2,2*(X[i]+1),2*(N[i] -X[i]))/(2*(N[i]-X[i]) +
(2*(X[i]+1))*qf(1-alpha/2,2*(X[i]+1),2*(N[i]-X[i])))
  }
 }
width <-pU -pL
plot(x=N,y=width,main="Confidence Interval Width by Sample Size:
Proportions",xlab="Sample Size",ylab="width = pU-pL")

text(x=200,y=0.20,"p ~= 0.10")
df2 <-data.frame(X,N,p,pL,pU,width)
write.csv(df2,"binomial confidence limits.csv")
```

FIGURE 15.3
Clopper–Pearson confidence interval widths.

on the value of the mean; rather, they depend on the standard deviation. So, in order to determine the width of a confidence interval for proportions, knowledge of the proportion as well as sample size is required. Figure 15.3 shows some R code for computing confidence interval widths with different sample sizes, and a proportion of approximately 0.10 (10 percent). Figure 15.4 shows a plot of 95-percent confidence interval widths for those sample sizes.

Pseudo-Replicates

Is there a difference between observing a behavior in 1,000 aardvarks and observing one aardvark 1,000 times? Hurlbert (1984, p. 187) defined pseudo-replication

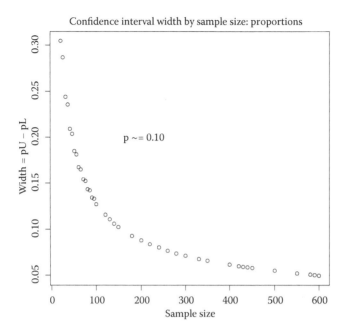

FIGURE 15.4
Clopper–Pearson interval width plot.

as performing statistical inference "…where either treatments are not replicated (though samples may be) or replicates are not statistically independent." Perhaps a treatment is applied once to a group of individuals simultaneously, but each individual within the group provides a response. We would never recommend using a single individual to provide all the data for a study. However, in some cases it may be reasonable to use all the data from each individual in a group even though the treatment was applied once to the entire group. One possible way to decide if multiple observations of individuals within a group is good enough is to perform a limited experiment. Suppose the treatment could be applied twice, once to each of two groups. Compare standard deviations of responses within groups, and standard deviation between the two groups. The ratio

$$F = \frac{S_B^2}{S_W^2}$$

where S_B^2 is the sample variance between groups (B is for "between") and S_W^2 is the sample variance within groups individual (W is for "within") can be used to test whether the variability between groups is greater than the variability within groups. If there are n_B ($= 2$) groups, and n_W observations within a single group, then the statistic F has an F distribution with n_{W-1} and n_{B-1} degrees of freedom in the numerator and denominator, respectively. Thus, a variance components analysis can be performed to determine whether the between-group variation is significantly greater than the within-group variation. See Chapter 6 for partitioning variance into between and within group components.

Too Many *p*-Values: False Discovery Rate

In Chapter 5, Tukey's Honestly Significant Difference (HSD) was presented as a multiple comparison adjustment, used to compare all pairs of levels of a factor found to be significant in an ANOVA. In some work, especially when analyszing very large datasets, many *p*-values might be computed, thus inflating the chance of Type I error (rejecting the null when in fact it was correct), and Tukey's method may not apply, if pairwise comparisons are not involved. For example, suppose that the same factors may affect many different responses. One method for adjusting the critical level with very large numbers of *p*-values is called the False Discovery Rate (FDR) adjustment, developed by Benjamini and Hochberg (1995). Briefly, it works like this:

Suppose you have N *p*-values. Here is the False Detection Rate (FDR) adjustment procedure for picking a critical value:

1. Pick a proportion, q (a decimal number between zero and one, not including zero or one).
2. Sort the *p*-values from smallest to largest: $p(1), p(2), ...,p(N)$.
3. Calculate the values: $\dfrac{i*q}{N}$ for each $i = 1$ to N.
4. Find the largest value of i, call it k, where $p(i) < \dfrac{i*q}{N}$.
5. The value $p(k)$ is the critical value. If $p \leq p(k)$, then reject the null hypothesis.

As a simple example, consider:

$$N = 10 \ p\text{-values}$$

as shown in Table 15.2 (sorted from smallest to largest). Let $q = 0.20$, and compute the $\dfrac{i*q}{N}$:

TABLE 15.2

$N = 10$ *p*-Values Sorted from Smallest to Largest

i	p(i)	i*q/N = i*0.2/10
1	0.03	0.02
2	0.04	0.04
3	**0.04**	**0.06**
4	0.10	0.08
5	0.11	0.10
6	0.13	0.12
7	0.14	0.14
8	0.17	0.16
9	0.20	0.18
10	0.25	0.20

$$1*0.20/10, \ 2*0.2/10, \ 3*0.2/10, \ 4*0.2/10, \ 5*0.2/10,$$
$$6*0.2/10, \ 7*0.2/10, \ 8*0.2/10, \ 9*0.2/10, \ 10*0.2/10$$

The largest *p*-value that is smaller than its corresponding value of $i*q/N$ is $p(3) = 0.04$. Therefore, the critical value for *p*-values is 0.04. Reject any null hypothesis in the group with a *p*-value less than or equal to 0.04.

The choice of q is arbitrary and subjective. Then again, choosing 0.05 as the critical value is also arbitrary and subjective.

A modification of the FDR adjustment (Yekutieli and Benjamini, 1999) involves computing adjusted *p*-values instead of determining the signifi-cance level. The procedure is

1. Sort the *p*-values, $p(i)$, from smallest to largest.

2. Compute $q(i) = \dfrac{p(i)*N}{i}$.

3. Compute $q*(i) = \min \{q(j), j = i, N\} \ \forall i = 1, N$.

4. Compute adjusted *p*-values: $p_{fdr}(i) = \dfrac{i}{N} q*(i)$.

In this modification, the arbitrary value q is replaced by a function $q*(i)$. The R function *p.adjust()* implements this procedure (and others). See the R code in Figure 15.5.

There is some disagreement about whether a multiple comparison adjust-ment is required for multiple responses when each is analyzed individually for a single experiment. Whether to use such an adjustment has no correct answer. In any case, the FDR adjustment is well suited to selecting the critical value for a large number of *p*-values.

```
setwd("C:\\Users\\SMFSBE\\Statistical Data & Programs")

df1 <-read.csv("20170829 Example 15.3 FDR.csv")
#
#Variable:
# p.i-vector of p-values
#
attach(df1)

p.adjust(p=p.i,method="fdr")# same as method = "BH"
```

FIGURE 15.5
p-Value adjustment code.

Key Points for Chapter 15

- Study design consists of both procedural and statistical considerations.
- Studies with multiple factors/effects can have total degrees of freedom partitioned, and degrees of freedom for error (dfe) can be computed, before any data are gathered.
- The dfe is the key to power, or the ability to detect effects.
- The noncentrality parameter is in part a function of the total sample size and the degrees of freedom for the factor/effect of interest.
- Confidence interval widths are largely a function of sample size.

Exercises and Questions

1. Suppose that instead of 20 individuals in the ANOVA example, there were only 15. Construct the power curve.
2. Compute a confidence interval width curve for proportions, where the estimated proportion is 0.80 (80 percent).

16

When Things Go Wrong …

In data-gathering exercises, things often do not go exactly according to plan. In some cases, the problems are insurmountable, and the entire experiment must be completely redone. In most cases, the experiment can be at least partially salvaged, if not completely saved.

Inadequate Measurement System

It is possible that after at least some data have been gathered, the researcher discovers that his or her system of measurement is simply too coarse. That is, it may be that the variation in the data results in at most an ordinal variable, with a small number of distinct categories. Thus, any analyses planned for a continuous variable are most likely not valid. If the data are actually binary, then analyses appropriate for binomial variables, such as logistic regression, may be employed. If the data have three or more values observed for the response variable, techniques appropriate for polytomous data may be appropriate, such as R × C contingency tables, or even Poisson regression.

It is also possible that while the measurements do appear to be continuous, the variability within an individual is too great relative to meaningful differences. In such a case, it is possible that averaging replicates or pseudo-replicates, and performing the inferential analyses on the averages rather than the individual measurements, may sufficiently reduce the within-individual noise. That is, replace all the replicated measurements made on an individual with their average for that individual.

Incorrect Assignment of Individuals to Groups

Sometimes an individual is planned to be assigned a certain treatment, or to have treatments applied in a particular order, but the individual is mislabeled or in some way assigned to the wrong group. If there is no structure within the groups (e.g., treatment order does not matter), then simply switch the individual's assignment. However, it is not advisable to change the group structure for a new individual to somehow compensate for the lack of balance. For example, in the 4×4 Latin square:

A B C D

B C D A

C D A B

D A B C

If an individual was originally assigned to A B C D, but instead was treated in the order B C D A, then it would make sense to take an (as yet untreated) individual originally assigned to B C D A and reassign the individual to A B C D. However, it would *not* be a good practice to change the order for some other individual to, say, A B D C. While such a reordering would compensate for balance between treatments A and order 1, and B and order 2, it would induce imbalance for treatments C and D.

An Undiscovered Covariate

It is possible that somewhere in the middle of executing an experiment, it is discovered that some covariate had been actively changing, but it had not been considered in the planning of the experiment. When such a discovery is made in such a way that the covariate values had in fact been recorded for all the previously observed individuals, then it is possible to simply include the covariate in the linear model. However, if the covariate was not measured and recorded for some of the data, then it cannot be incorporated in the analyses of the total data set, regardless of its potential effects or importance. In some cases, it may be possible to perform a separate set of analyses, first for those subjects where the covariate was not recorded, and then for those for which the covariate can be included in the model.

Unintended Order Effects

It is possible that treatment order was not thought a priori to have a significant effect on the response, and consequently the order of treatment was identical for all subjects. In this case, treatment is said to be completely aliased (Montgomery, 2001) with order. The only thing that can be done is a limited, additional experiment to determine if order in fact has an effect.

Even a single replicate of a Latin square experiment could be used to examine such potential effects. Unfortunately, failure to reject the null hypothesis of no order effect does not prove that there is no such effect. However, it may be possible to construct confidence intervals for the differences in response between orders, using techniques such as Tukey's HSD. If none of these confidence intervals show a biologically meaningful difference, then the experimenter has some rationale for believing that order was not meaningful. Of course, this depends on the ability to identify the meaning of meaningful, which may not be possible.

Missing Data

Perhaps the most common problem is that of missing data. Missing data can range in consequences from virtually nothing to catastrophic. If, for example, there are multiple replicates in each group, and one replicate value is missing (due to any circumstance), well, the analysis will proceed as planned, with a slightly diminished power. On the other hand, if the response observed under one particular treatment is missing from all individuals, then certainly no inference can be made about that treatment, but those missing data do not preclude making inferences about the other treatments. Even if the treatments are applied as a Latin square, the order effect may not be completely obliterated. Consider a 3 × 3 Latin square:

A B C
B C A
C A B

Suppose that all the data for treatment C are lost. Then the square becomes a rectangle:

A B _
B _ A
_ A B

At the very least, the ordinality is preserved, namely, in the first group A comes before B, and in the second group, B is applied before A. In the third group, A is still applied before B, which means there is a lack of balance. Nevertheless, the analysis may proceed, but the inference about order must be interpreted as relative order, as opposed to literal order.

In order to analyze such data, the *anova()* function should be used together with the *lm()* function, instead of the *Anova()* function.

As an example of how one might compensate for the loss of all data from one treatment, consider the code in Figure 16.1, together with the data in "20170723 Missing Data for One Treatment.csv." Since the response data in this example were simply random, the only thing of interest is the column "Df," the degrees of freedom for each term. There were originally a total of 27 observations, but 9 values were lost (all treatment C observations). So now there are $18 - 1 = 17$ total degrees of freedom. There were three groups, even though two groups had the same order. That is, there were only two orders (A-B or B-A), so there are $2 - 1$ degrees of freedom for order, and only two treatments, so $2 - 1$ degrees of freedom for treatments. As to individuals, we might have thought there were $3*(3 - 1) = 6$ degrees of freedom for individuals, but R gave them seven. The residual degrees of freedom are reported to be seven. Using our previously described methods (see Chapter 5), we compute $17 - 2 - 1 - 1 - 6 = 7$ degrees of freedom for the residuals (error). However, R computes the residual degrees of freedom as $18 - 2 - 1 - 1 - 7 = 7$. So R chooses to compute degrees of freedom by subtracting from the total sample size, and adds 1 degree of freedom back to the nested factor. If we had ignored Group, then R would have computed the degrees of freedom differently. Since there were nine individuals, there would be $9 - 1 = 8$ degrees of freedom for individuals. There would still be $2 - 1 = 1$ degree of freedom for order and $2 - 1 = 1$ degree of freedom for treatment. R would

```
setwd("C:\\Users\\SMFSBE\\Statistical Data & Programs")
df1 <-read.csv("20170723 Example 16.1 Missing Data for One Treatment.csv")
#library(car) if Anova() function is used
#
#
# VARIABLES:
# Group
# Grpoup2
# Individual
# Order
# Order2
# Treatment
# Response
#
attach(df1)
fGroup <-factor(Group2)
fIndividual <-factor(Individual)
fOrder <-factor(Order2)

LS3x3 <-lm(Response ~ fGroup + fIndividual%in%fGroup + fOrder + Treatment,
na.action=na.omit)


anova(LS3x3)


OUTPUT

> anova(LS3x3)
Analysis of Variance Table

Response: Response
```

	Df	Sum Sq	Mean Sq	F value	Pr(>F)
fGroup	2	1.3280	0.6640	0.3555	0.7128
fOrder	1	3.2768	3.2768	1.7546	0.2269
Treatment	1	0.0256	0.0256	0.0137	0.9101
fGroup:fIndividual	7	9.0146	1.2878	0.6895	0.6820
Residuals	7	13.0732	1.8676		

FIGURE 16.1
ANOVA with missing data for one treatment.

have used 18 − 1 = 17 total degrees of freedom, so dfe would have been 17 − 8 − 1 − 1 = 7. The experimenter should at least be aware that R will handle nested effects somewhat differently than other software.

Imputation

It is a rare study in which values for all variables are obtained for all individuals. Missing data can be compensated for in many ways. The foremost method is to ignore the missing observations. For example, if $n = 10$ individuals were sampled, and variable X was only observed for nine of those individuals, one might simply compute the average value of X for the nine observed values. This is a perfectly rational approach, if all the researcher wanted to do was estimate the mean of X. The exclusion of individuals with missing values has been referred to as *listwise deletion* (Van Buuren et al., 2012). However, suppose that the researcher wants to compare the average value of X between multiple experimental treatments performed on 10 individuals. Furthermore, suppose that for one treatment, only 8 of the 10 individuals had observations of X, and for another there were nine, and for a third there were seven. While it is entirely possible to compute a one-way ANOVA with unbalanced data, individuals would act as blocks, and some of those blocks will be incomplete in an unbalanced fashion. With such missing data, the ANOVA will produce only approximate p-values. While it is possible to delete all individuals for which any results are missing, that deletion will reduce the sample size, and therefore reduce the power of the test. The impetus for replacing missing values may be even greater in the cases of multivariate analyses, where muiltiple response variables have some correlation structure.

It is possible to "create" values of X for those individuals where X was not observed. The processes and methods of performing such insertions of values to compensate for missing data are collectively referred to as imputation (Van Buuren, 2012). The methods range from fairly simple, such as using the mean value within groups to replace missing values, to fairly complex, such as employing Bayesian algorithms. Rubin (1976) constructed a taxonomy of missing data. He classified missing values into three major groupings:

MCAR: Missing Completely At Random

MAR: Missing At Random

MNAR: Missing Not at Random

MCAR data are those values that are missing without any relation or correlation to any of the other variables in the experiment/study. Furthermore, it is presumed that the likelihood a value is potentially MCAR is equal for all values in the study. MAR data are those values missing either due to, or at least conditional upon, the value of other variables, and therefore having the probability of missing observations be affected by other variables. This means that the likelihood a value missing may be different for different groups, or for different conditions under which the data were gathered. Finally, MNAR are variables whose likelihood of missing are related to some effects, but those effects are completely unknown. The difference between MCAR and MNAR is that it is presumed that the likelihood of missing values for MCAR data is not related to anything, known or unknown, whereas the likelihood for missing values of the MNAR type is affected by some unknown factor. The determination of whether missing values are MCAR or MNAR is largely one of assumption.

Imputation can be categorized as single or multiple (Fleiss et al., 2003). Single imputation involves replacing missing values with an expected value. Fleiss provided formulas for singly imputing values in a 2×2 contingency table. Suppose that R represents the binary row valiable, and C the binary column variable. That is, the values or states of each of these variables can be represented by the numbers zero or one. Let the numbers of observed values be a_o, b_o, c_o, and d_o, and illustrated in Table 16.1. Furthermore, let m1 be the number of indiduals for which values of C are missing when $R = 1$.

Suppose that there were individuals for whom the column variable was unknown given that the row variable was equal to "1." This would make the missing values of the MAR type. Thus the numbers of individuals with $R = 1$ may be underinflated. The single imputation formulas for replacing a_o and b_o are

$$a_T = a_o \left(1 + \frac{m_1}{o_1} \right)$$

and

$$b_T = b_o \left(1 + \frac{m_1}{o_1} \right)$$

In other words, a_T and b_T are the guesses for the "total" numbers of indivuals with $C = 1$ and $C = 0$, respectively, when $R = 1$.

TABLE 16.1

A 2×2 Table, Observed Numbers

	C = 1	C = 0	
R = 1	a_o	b_o	$o_1 = a_o + b_o$
R = 0	c_o	d_o	$o_2 = c_o + d_o$
	$c_1 = a_o + c_o$	$c_2 = b_o + d_o$	

Multiple imputation is the process by which missing values are replaced by simulating a sample of values from a model based on the non-missing values, and then performing some kind of averaging process for the test statistic or parameter estimates. Multiple imputation therefore requires more information beyond the expected value of the distribution of the population that was sampled. There are many such multiple imputation methods, using bootstrapping, Bayesian, or other procedures. The R package called *MICE*, an acronym for multiple imputation by chained equations, contains functions for implementing multiple imputation procedures. Multiple imputation is such a vast topic that here we will only mention its existence, and refer the reader to the Van Buuren (2012) text.

Summary

There may be an infinite number of ways things can go wrong in a given experiment. That implies that the likelihood of at least one thing going wrong in any experiment approaches one. Several classes of problems have been listed in this chapter, some of which have solutions, and others do not (other than starting all over). Hopefully, being aware of some common problems will help the researcher avoid those problems through careful planning and execution.

Key Points for Chapter 16

- In virtually every empirical investigation, unexpected problems occur.
- In many cases, there are remediations that can be implemented, and these remediations may save the data.
- Sometimes remediation is in the form of getting more replicates and averaging, as in the case of inadequate measurement systems.
- Sometimes a small, additional experiment is required to potentially eliminate some aliased effect, such as order of treatment.
- Sometimes "listwise deletion" of conditions or individuals is required; such deletions may reduce the ability to make some inferences, but may at least allow for others to be made.
- Imputation is a set of techniques that may be used to "replace" missing values, allowing for a complete set of planned inferences, without reducing power.

Exercises and Questions

1. In a study, 20 individuals were exposed to three different stimuli, and reaction time was recorded. For four of the individuals, no recording was made for stimulus #2. What type of missing data does this comprise? Explain your answer.

2. Do you think it is less important to replace MCAR results compared to MAR results? Explain your answer.

Appendix A: Matrices and Vectors

A matrix is a rectangular array of numbers:

$$
M = \begin{bmatrix} a_{11} & a_{12} & \cdots & a_{1n} \\ \vdots & \vdots & \ddots & \vdots \\ a_{m1} & a_{m2} & \cdots & a_{mn} \end{bmatrix}
$$

The "entries," a_{ij}, can be real or complex numbers, but we will only deal with real matrices and vectors.

A vector is a special matrix, that has either only one row or one column. For the most part, we will consider vectors to be a column of numbers:

$$
v = \begin{bmatrix} b_1 \\ \vdots \\ b_m \end{bmatrix}
$$

We will represent the names of matrices with boldface capital letters, and the names of vectors with boldface lowercase letters. We will use the convention of labeling the entries, or components, with subscripts that represent the row number (first subscript) and column number (second subscript). In the case of vectors, the single subscript will refer to the row number for "column" vectors or the column number for "row" vectors. As we stated before, mostly when we refer to a vector, we will be thinking of it as a column. When we need to refer to a "row" vector, most of the time the name of the vector will have a superscript "T," which is an abbreviation for "transpose." A transpose vector or matrix is one where the rows become columns.

There are several operations involving matrices and vectors that will appear throughout the text. They are

1. Scalar multiplication
2. Matrix and vector addition
3. Transposition
4. Matrix multiplication
5. "Dot" or scalar product of two vectors, and vector length
6. Square matrices, the identity matrix, and matrix inverses
7. Determinants

8. Eigenvalues and eigenvectors
9. Diagonalization and powers of matrices

A.1 Scalar Multiplication

A scalar for our purposes is a number that is not a vector or matrix. In fact, the entries or components in all of our vectors and matrices will be scalars. To multiply a scalar by a matrix of vector, simply multiply that scalar by every entry in the matrix or vector. So

$$A = sM = Ms = \begin{bmatrix} sa_{11} & sa_{12} & \cdots & sa_{1n} \\ \vdots & \vdots & \ddots & \vdots \\ sa_{m1} & sa_{m2} & \cdots & sa_{mn} \end{bmatrix}$$

or

$$w = sv = vs = \begin{bmatrix} sb_1 \\ \vdots \\ sb_m \end{bmatrix}$$

The order is not important for scalar multiplication. This is not the case of matrix multiplication.

A.2 Matrix and Vector Addition

Adding matrices and vectors is componentwise addition. This means that in order to add two matrices, they must have exactly the same number of rows and columns. So

$$M = \begin{bmatrix} a_{11} & a_{12} & \cdots & a_{1n} \\ \vdots & \vdots & \ddots & \vdots \\ a_{m1} & a_{m2} & \cdots & a_{mn} \end{bmatrix}$$

$$N = \begin{bmatrix} b_{11} & b_{12} & \cdots & b_{1n} \\ \vdots & \vdots & \ddots & \vdots \\ b_{m1} & b_{m2} & \cdots & b_{mn} \end{bmatrix}$$

$$M+N=N+M=\begin{bmatrix} a_{11}+b_{11} & a_{12}+b_{12} & \cdots & a_{1n}+b_{1n} \\ \vdots & \vdots & \ddots & \vdots \\ a_{m1}+b_{m1} & a_{m2}+b_{m2} & \cdots & a_{mn}+b_{mn} \end{bmatrix}$$

and

$$v=\begin{bmatrix} a_1 \\ \vdots \\ a_m \end{bmatrix}$$

$$w=\begin{bmatrix} b_1 \\ \vdots \\ b_m \end{bmatrix}$$

$$v+w=w+v=\begin{bmatrix} a_1+b_1 \\ \vdots \\ a_m+b_m \end{bmatrix}$$

A.3 Transposition

If

$$M=\begin{bmatrix} a_{11} & a_{12} & \cdots & a_{1n} \\ \vdots & \vdots & \ddots & \vdots \\ a_{m1} & a_{m2} & \cdots & a_{mn} \end{bmatrix}$$

Then the transpose of matrix M, call it M^T, is

$$M^T=\begin{bmatrix} a_{11} & \cdots & a_{m1} \\ a_{12} & \cdots & a_{m2} \\ \vdots & \ddots & \vdots \\ a_{1n} & \cdots & a_{mn} \end{bmatrix}$$

To illustrate the point, the subscripts on the entries in M^T have not been changed from the original, although our convention is to have the row number first followed by the column number. As a more concrete example, if

$$M = \begin{bmatrix} 1 & 2 & 3 & 4 \\ 5 & 6 & 7 & 8 \\ 9 & 10 & 11 & 12 \end{bmatrix}$$

then its transpose would be

$$M^T = \begin{bmatrix} 1 & 5 & 9 \\ 2 & 6 & 10 \\ 3 & 7 & 11 \\ 4 & 8 & 12 \end{bmatrix}$$

A transposed column vector becomes a row vector, and vice versa:

$$v = \begin{bmatrix} a_1 \\ \vdots \\ a_m \end{bmatrix}$$

$$v^T = \begin{bmatrix} a_1 & \cdots & a_m \end{bmatrix}$$

A.4 Matrix Multiplication

Not all pairs of matrices can be multiplied. Furthermore, matrix multiplication is not commutative, as was matrix addition. In order for matrix M and N to be "multipliable" (usually called *conformable* with respect to multiplication), to obtain the product MN, the number of columns of M must equal the number of rows in matrix N. Symbolize the entries of matrix M as a_{ij}, and the entries of matrix N as b_{jk}. If M has m rows and n columns, and N has n rows and r columns, then the product matrix

$$P = MN$$

will have m rows and r columns. The ith row and j^{th} column of P will be the sum:

$$p_{ij} = a_{i1}^* b_{1j} + a_{i2}^* b_{2j} + \ldots + a_{in}^* b_{nj}$$

As an example, consider:

$$M = \begin{bmatrix} 1 & 2 \\ 3 & 4 \end{bmatrix}$$

$$N = \begin{bmatrix} 5 & 6 & 7 \\ 8 & 9 & 10 \end{bmatrix}$$

Then:

$$P = MN = \begin{bmatrix} 1*5+2*8 & 1*6+2*9 & 1*7+2*10 \\ 3*5+4*8 & 3*6+4*9 & 3*7+4*10 \end{bmatrix} = \begin{bmatrix} 21 & 24 & 27 \\ 47 & 54 & 61 \end{bmatrix}$$

The product NM is not defined.

A very important special case of matrix multiplication is between an $m \times n$ matrix and an $n \times 1$ vector (which is a very special kind of matrix). That is:

$$Ax = \begin{bmatrix} a_{11} & a_{12} & \cdots & a_{1n} \\ \vdots & \cdots & \ddots & \vdots \\ a_{m1} & a_{m2} & \cdots & a_{mn} \end{bmatrix} \begin{bmatrix} x_1 \\ x_2 \\ \vdots \\ x_n \end{bmatrix} = \begin{bmatrix} b_1 \\ \vdots \\ b_m \end{bmatrix} = b$$

A.5 Dot or Scalar Product

Two vectors can be multiplied in a fashion that yields a scalar. In order to obtain this scalar product, one vector must be a row vector and the other a column vector, and they both must have equal numbers of components. The "dot" product is symbolized with a "•". If

$$v^T = [\ a_1 \quad \cdots \quad a_m\ ]$$

$$w = \begin{bmatrix} b_1 \\ \vdots \\ b_n \end{bmatrix}$$

then the dot product $v^T \bullet w$ is

$$v^T \bullet w = \begin{bmatrix} a_1 & \cdots & a_n \end{bmatrix} \begin{bmatrix} b_1 \\ \vdots \\ b_n \end{bmatrix} = a_1 * b_1 + \ldots + a_n * b_n$$

On the other hand, the product wv^T is actually a matrix. Suppose that

$$w = \begin{bmatrix} b_1 \\ \vdots \\ b_m \end{bmatrix}$$

with $m \neq n$.

The vector w can be thought of as a matrix with m rows and one column, and v^T can be thought of as a matrix with one row and n columns. Thus, the product wv^T is a matrix with m rows and n columns.

A vector has length, which we will designate as $\lVert v \rVert$, and is defined to be

$$\lVert v \rVert = \sqrt{\sum_{i=1}^{n} a_i^2} = (v^T \bullet v)^{1/2}$$

In as much as this length is a scalar, another vector that has length = 1 can be constructed as

$$\tilde{v} = \frac{1}{\lVert v \rVert} v$$

A vector of all zeros has length = 0.

A.6 Square Matrices, the Identity Matrix, and Matrix Inverses

A square matrix is one that has the same number of rows and columns. They turn out to be a very important case for a wide variety of applications. For one thing, the product of a matrix and its transpose is always a square matrix. If M is an $m \times n$ matrix, then its transpose, M^T, is an $n \times m$ matrix, and

$$S_1 = M^T M$$

is a square matrix with dimensions $n \times n$, and

$$S_2 = MM^T$$

is a square matrix with dimensions $m \times m$.

So, going back to our least-squares regression, the matrix Z is n × k, so that Z^T is k × n, and therefore Z^TZ is a k × k square matrix.

An identity matrix is a special square matrix where most of the entries are zero, except for the diagonal entries (a_{ii}), which are all one. In general, the $n \times n$ matrix, I_n is

$$I_n = \begin{bmatrix} 1 & 0 & \cdots & 0 \\ 0 & 1 & \cdots & 0 \\ \vdots & 0 & \ddots & \vdots \\ 0 & 0 & \cdots & 1 \end{bmatrix}$$

It turns out that for any n × n square matrix, S_n, the products with the identity matrix of corresponding dimensions is the matrix itself:

$$S_nI = IS_n = S_n$$

This begs the question: Can we find another matrix so that its product with square matrix S_n is the $n \times n$ identity? The answer is, sometimes. This matrix, if it exists, is called the inverse of S_n. When a square matrix has an inverse, which we denote with a superscript "–1", it has the property that

$$S_nS_n^{-1} = S_n^{-1}S_n = I_n$$

In the case of multiple regression as described earlier, the answer to the existence of the inverse matrix $[Z^TZ]^{-1}$ is a resounding "almost always." The usual case for regression is that the matrix

$$S_k = Z^TZ$$

does in fact have an inverse matrix:

$$S_k^{-1} = [Z^TZ]^{-1}$$

Is it possible to know whether this matrix exists? Generally, software will tell you if you try to fit a multiple regression and the matrix Z^TZ is not invertible. This condition will occur if any column of the regressor matrix, Z, is in fact a linear function of other columns. Such a condition is called *collinearity*.

Sometimes a column is very close to being a linear function of other columns, but not quite. Such a condition is called *ill conditioning*. Ill-conditioned matrices may have an inverse, but it may be too hard or at least very difficult to compute it (we have not discussed how to actually find an inverse at this point; suffice it to say that there is a numerical procedure for doing so). As an example, consider the data in Table A1.1. In this case, X1 is just the numbers from 1 to 9, and Y = 2*X1 + 5 exactly. In the R code shown in Figure A1.1, the variable X2 is X1 plus a little noise (rnorm generates random normal variable values).

The output is shown in Figure A1.2.

TABLE A1.1

Perfect Linear
Relationship Example

X1	Y
1	7
2	9
3	11
4	13
5	15
6	17
7	19
8	21
9	23

```
setwd("C:\\Users\\Statistical Methods for Animal Behavior Field Studies\\Statistical Data &
Programs")
df1 <-read.csv("20161014 Example 4.1.6.1 Collinearity.csv")
#
#
# VARIABLES:
# X1
# Y
# Note: Y = 2*X1 + 5
#
attach(df1)
X2 <-X1 + rnorm(length(X1),0,0.1)
X3 <-3*X2 + 7 #inducing collinearity

linreg1 <-lm(Y ~ X2)
linreg2 <-lm(Y ~ X2 + X3)
#
# Make some plots:
#
# dev.new() allows more plots to be made without overwriting previous plots
#
plot(X2,Y)
dev.new()
plot(X3,Y)
dev.new()
plot(X2,X3)

summary(linreg1)
summary(linreg2)
```

FIGURE A1.1
Code for inducing collinearity and performing multiple linear regression.

```
> summary(linreg1)

Call:
lm(formula = Y ~ X2)

Residuals:
   Min      1Q   Median      3Q     Max
-0.30438 -0.19716 -0.14869  0.08547  0.45660

Coefficients:
            Estimate Std. Error t value Pr(>|t|)
(Intercept)  4.99891    0.21136   23.65 6.14e-08 ***
X2           1.99315    0.03744   53.24 2.16e-10 ***
---
Signif. codes:  0 '***' 0.001 '**' 0.01 '*' 0.05 '.' 0.1 ' ' 1

Residual standard error: 0.2906 on 7 degrees of freedom
Multiple R-squared:  0.9975,    Adjusted R-squared:  0.9972
F-statistic:  2834 on 1 and 7 DF,  p-value: 2.162e-10

> summary(linreg2)

Call:
lm(formula = Y ~ X2 + X3)

Residuals:
   Min      1Q   Median      3Q     Max
-0.30438 -0.19716 -0.14869  0.08547  0.45660

Coefficients: (1 not defined because of singularities)
            Estimate Std. Error t value Pr(>|t|)
(Intercept)  4.99891    0.21136   23.65 6.14e-08 ***
X2           1.99315    0.03744   53.24 2.16e -10 ***
X3                NA         NA      NA       NA
---
Signif. codes:  0 '***' 0.001 '**' 0.01 '*' 0.05 '.' 0.1 ' ' 1

Residual standard error: 0.2906 on 7 degrees of freedom
Multiple R-squared:  0.9975,    Adjusted R-squared:  0.9972
F-statistic:  2834 on 1 and 7 DF,  p-value: 2.162e-10
```

FIGURE A1.2
Regression output for collinear example.

The first regression (linreg1) was perfectly fine. The intercept was close to 5 and the slope parameter (coefficient for X2) was close to 2. Remember that we added some random noise, so that the fit would not be perfect.

The second regression had the collinear term X3 included. The *lm* function could not estimate its associated coefficient. The message "(1 not defined because of singularities)" indicates that there is some collinearity with respect to variable X3, hence the term *singularities*.

As to how a square matrix inverse is computed, it is accomplished through some variant of a numerical procedure called Gauss–Jordan elimination. The details of this process may be found in any book on linear algebra (see, for example, Strang, 2016).

A.7 Determinants

Square matrices have a number associated with them called a determinant. Determinants in general are computed using the Gauss–Jordan elimination process, and are in fact related to a square matrix's inverse. Perhaps the most important property of the determinant is that if it is zero, then the associated square matrix does not have an inverse. A determinant can either be negative, 0, or positive. The general definition of the process by which a determinant is computed is beyond the scope of this discussion. It is important to realize that such a quantity exists. The determinant is often symbolized with either the functional notation det(S) or "absolute value" bars surrounding the name of the square matrix. For example:

$$det(S) = |S|$$

However, a determinant can in fact be negative.

R has a built-in determinant function, det(.), where the argument is a square matrix. Consider this example:

```
> col1 <- c(1,2,3)
> col2 <- c(4,5,6)
> col3 <- c(9,7,8)
> mat1 <- cbind(col1,col2,col3)
> det(mat1)
[1] -9
```

Although no simple formula exists for determinants in general, a 2 × 2 square matrix has a fairly simple determinant formula. For a 2 × 2 matrix:

$$A = \begin{bmatrix} a & b \\ c & d \end{bmatrix}$$

the determinant is

$$det(A) = ad - bc$$

Another special case of determinant is when the square matrix is symmetric across its diagonal. That is, if a_{ij} represents the element of the matrix in the ith row and jth column, then the matrix is symmetric, also referred to as triangular, if $a_{ij} = a_{ji}$, $i \neq j$. Then the determinant is equal to the product of the diagonal elements:

$$det(A) = \prod_{i=1}^{n} a_{ii}$$

The matrix A is $n \times n$. This fact is most useful for a special matrix, called the *variance–covariance matrix*, or sometimes called the *covariance matrix*. If X_1

and X_2 are random variables with means μ_1 and μ_2, respectively, then the covariance of X_1 and X_2 is

$$Cov(X_1, X_2) = Cov(X_2, X_1) = \sigma_{12} = \sigma_{21}$$

$$= \int_{-\infty}^{+\infty}\int_{-\infty}^{+\infty} (x_1 - \mu_1)(x_2 - \mu_2) f(x_1, x_2) dx_1 dx_2$$

The covariance between two variables X_1 and X_2 could be estimated from n observations with the formula:

$$C(X_1, X_2) = \frac{1}{n}\sum_{i=1}^{n}(x_{1i} - \bar{x}_1)(x_{2i} - \bar{x}_2)$$

where $\bar{x}$ is the arithmetic average.

The diagonal of the covariance matrix is just the variances of the respective variables. So, for example, the covariance matrix of X_1 and X_2 could be symbolized as

$$\Sigma = \begin{bmatrix} \sigma_1^2 & \sigma_{12} \\ \sigma_{21} & \sigma_2^2 \end{bmatrix}$$

It is very likely that there will never be a need to compute a determinant per se. Its most obvious application in the area of multivariate analysis, that is, situations where there are multiple "response" variables that are related to each other in some fashion. Many of the multivariate methods involve something called eigenvalues and eigenvectors, which are computed using equations with determinants.

A.8 Eigenvalues and Eigenvectors

Suppose A is a square matrix of dimensions $n \times n$:

$$A = \begin{bmatrix} a_{11} & \cdots & a_{1n} \\ \vdots & \ddots & \vdots \\ a_{n1} & \cdots & a_{nn} \end{bmatrix}$$

Sometimes, especially when the components of matrix A are pairwise correlations between a set of response variables, it is of interest to find a scalar, λ, such that

$$Ax = \lambda x$$

Since multiplying the identity matrix by a vector yields the original vector:

$$Ix = x$$

then the previous equation can be expressed as

$$Ax = \lambda Ix$$

Because the two sides of this equation both have the same dimensions ($n \times 1$), the right-hand side can be subtracted from the left:

$$Ax - \lambda Ix = 0$$

where $\mathbf{0}$ is an $n \times 1$ vector of all zeros. The vector x can be "factored" to yield:

$$(A - \lambda I)x = 0$$

It turns out that the equation only has a solution if

$$|A - \lambda I| = |\mathbf{0}| = 0$$

The determinant of the $n \times 1$ vector of all zeros is zero.

It turns out that there are potentially n different values of λ that satisfy the determinant equation, called the characteristic equation for matrix A. That is, the determinant $|A - \lambda I|$ actually is a polynomial of order n, so setting it equal to 0 allows for the possibility of n solutions, or roots. The roots of the characteristic equation are called, believe it or not, characteristic roots, or characteristic values, or eigenvalues. Each root, λ_i then corresponds to a vector, x_i, that satisfies the equation:

$$Ax_i = \lambda_i x_i$$

The solution vectors, x_i, i = 1, n, are called characteristic vectors, or eigenvectors.

A.9 Diagonalization and Powers of Matrices

A diagonal matrix is a square ($n \times n$) matrix with diagonal entries, a_{ii} not necessarily zero, but all other entries necessarily 0 ($a_{ij} = 0$, $i \neq j$). Of particular interest is a diagonal matrix, Λ, whose diagonal entries are the eigenvalues of another $n \times n$ matrix, A. If matrix S is the matrix of eigenvectors of A, then it turns out that (Strang, 2016):

$$A = S\ \Lambda S^{-1}$$

```
setwd("C:\\Users\\SMFSBE\\Statistical Data & Programs")
df1 <-read.csv("20170220 Example A.9 4x4 Matrix.csv")
#
#
# VARIABLES:
# s1
# s2
# s3
# s4
#
attach(df1)
Amat <-as.matrix(df1)
Smat <-eigen(Amat)
Smat$values
Smat$vectors
Lamda <-diag(Smat$values)
Eig <-as.matrix(Smat$vectors)
EigInv <-solve(Eig)
Arecover <-Eig%*%Lamda%*%EigInv
Lamda
Eig
Amat
Arecover
OUTPUT

> Lamda
     [,1]       [,2]        [,3]          [,4]
[1,]   1  0.0000000 0.0000000  0.000000e+00
[2,] - 0  0.4422144 0.0000000  0.000000e+00
[3,]   0  0.0000000 0.1922144  0.000000e+00
[4,]   0  0.0000000 0.0000000 -1.289329e-17
> Eig
     [,1]       [,2]        [,3]         [,4]
[1,]  0.5  0.7026433 -0.1512474 -0.07142857
[2,]  0.5 -0.1993420  0.8144128  0.50000000
[3,]  0.5 -0.3825925  0.1353958 -0.78571429
[4,]  0.5 -0.5658429 -0.5436212  0.35714286
> Amat
      s1   s2   s3    s4
[1,] 0.1 0.2 0.30 0.40
[2,] 0.4 0.3 0.20 0.10
[3,] 0.5 0.2 0.15 0.15
[4,] 0.6 0.1 0.10 0.20
> Arecover
     [,1]    [,2] [,3] [,4]
[1,]  0.1   0.2 0.30 0.40
[2,]  0.4   0.3 0.20 0.10
[3,]  0.5   0.2 0.15 0.15
[4,]  0.6   0.1 0.10 0.20
>
```

FIGURE A1.3
Extracting eigenvalues and eigenvectors and diagonalization.

While there are many applications and mathematical reasons why this relation is important, we will focus here on one fact, relating to powers of a matrix. The kth power of matrix A is simply A multiplied by itself k times. It turns out that for a diagonal matrix, say our matrix of eigenvalues, Λ, the kth power is simply a diagonal matrix whose diagonal elements are a_{ii}^k. This is very convenient, inasmuch as it makes computing the kth power of A fairly easy:

$$A^k = S\Lambda^k S^{-1} = S \begin{bmatrix} a_{11}^k & \cdots & 0 \\ \vdots & \ddots & \vdots \\ 0 & \cdots & a_{nn}^k \end{bmatrix} S^{-1}$$

All but the first S and last S^{-1} will "cancel" each other, since $SS^{-1} = I$, the identity matrix.

The R function *eigen*() can be used to extract both eigenvalues and eigenvectors from a square matrix. Figure A1.3 shows some code and associated output for using *eigen*() with a 4×4 matrix. The code also shows how the original matrix can be "recovered" using the eigenvalues and eigenvectors. The function *diag*() is used to generate the diagonal matrix from the vector of eigenvalues.

Appendix B: Solving Your Problem

This text has offered a wide array of methods and models. Once you are faced with a particular problem, some subset of these will be applicable to your particular situation. Here is a roadmap for helping you navigate the sea of statistical methods described in the above chapters.

1. Getting a prediction: Continuous regressors and continuous responses (non-Bayesian, mostly)

   Chapter 4—The Linear Model: Continuous Regressor Variables

   Chapter 6—The Linear Model: Random Effects and Mixed Models

   Chapter 9—Multivariate Continuous Variables: Dimension Reduction, Clustering, Discrimination (when there are multiple, correlated, continuous responses)

   Chapter 12—Modern Prediction Methods and Machine Learning Models

2. Choosing a model: Linear or generalized linear models

   Chapter 4—The Linear Model: Continuous Regressor Variables

   Chapter 12—Modern Prediction Methods and Machine Learning Models

   Chapter 14—Time Series Analysis and Stochastic Processes

3. Assessing an effect: Discrete factors and continuous responses

   Chapter 3—Continuous Variables: One and Two Samples

   Chapter 5—The Linear Model: Discrete Regressor Variables

   Chapter 9—Multivariate Continuous Variables: Dimension Reduction, Clustering, Discrimination (when there are multiple, correlated, continuous responses)

4. Getting a prediction: Discrete responses

   Chapter 8—The Generalized Linear Model: Logistic and Poisson Regression

   Chapter 12—Modern Prediction Methods and Machine Learning Models (stepwise methods can be used with logistic or Poisson regression models)

5. Assessing an effect: Discrete factors and discrete responses

   Chapter 2—Binary Results: Single Samples and 2 × 2 Tables

   Chapter 7—Polytomous Discrete Variables: R × C Contingency Tables

6. Time-to-event prediction, with or without censoring

   Chapter 13—Time-to-Event Modeling

7. Predictions in time: Continuous or discrete responses varying in time

   Chapter 14—Time Series Analysis and Stochastic Processes

8. Bayesian modeling: When prior information is available

   Chapter 10—Bayesian and Frequentist Philosophies

9. Modeling and assessing the economics of animal decision-making

   Chapter 11—Decision and Game Theory

10. Designing a study

    Chapter 15—Study Design and Sample Size Considerations

11. Reviewing Statistical Foundations

    Chapter 1—Statistical Foundations

Hopefully you will read through the entire text at least once. Later, when you are faced with particular problems to solve, this roadmap may help you find the most relevant information.

No text is truly encyclopedic, especially one that is intended as a survey, as is this book. You may find methods described in literature that do not appear in this work. The reader is encouraged to seek other methods and approaches to problems. It is the hope of the authors that this text will help those just starting on the path of scientific inquiry, and provide a ready reference for those who have already embarked on that journey.

References

Ahnesjö, J., Forsman A. (2006). Differential habitat selection by pygmy grasshopper color morphs; interactive effects of temperature and predator avoidance. *Evolutionary Ecology* 20, pp. 235–257.

Akaike, H. (1974). A new look at the statistical model identification. *IEEE Transactions on Automatic Control* AC-19, pp. 716–723.

Albo, M.J., Bilde, T., Uhl, G. (2013). Sperm storage mediated by cryptic female choice for nuptial gifts. *Proceedings of the Royal Society B* 280:20131735.

Alcock, J. (2001). *Animal Behavior*. Sinauer Associates, Inc., Sunderland.

Anderson, T.W. (1958). *An Introduction to Multivariate Statistical Analysis*. John Wiley & Sons, New York.

Armitage, P. (1971). *Statistical Methods in Medical Research*. Blackwell Scientific Publications, London.

Art, J., Fligner, M.A., Notz, W.I. (1998). *Sampling and Statistical Methods for Behavioral Ecologists*. Cambridge University Press, Cambridge.

Benjamini, Y., Hochberg, Y. (1995). Controlling the false discovery rate: A practical and powerful approach to multiple testing. *Journal of the Royal Statistical Society B* 57(1):289–300.

Bleay, C., Comendant, T., Sinervo, B. (2007). An experimental test of frequency-dependent selection on male mating strategy in the field. *Proceedings of the Royal Society B* 274:2019–2025.

Box, G.E.P., Cox, D.R. (1964). An analysis of transformations. *Journal of the Royal Statistical Society B* 26:211–243.

Box, G.E.P., Jenkins, G.M., Reinsel, G.C. (2008). *Time Series Analysis: Forecasting and Control*, 4th ed. John Wiley & Sons, Hoboken.

Bronikowski, A.M., Altmann, J. (1996). Foraging in a variable environment: Weather patterns and the behavioral ecology of baboons. *Behavioral Ecology and Sociobiology* 39:11–25.

Chakerian, J., Holmes, S. (2012). Computational tools for evaluating phylogenetic and hierarchical clustering trees. *Journal of Computational and Graphical Statistics* 21(3):581–599.

Chiarle, A., Isaia, M. (2013). Signal complexity and modular organization of the courtship behaviours of two sibling species of wolf spiders (Araneae: Lycosidae). *Behavioral Processes* 97:33–40.

Clopper, C.J., Pearson, E.S. (1934). The use of confidence or fiducial limits illustrated in the case of the binomial. *Biometrika* 26:404–413.

Cochran, W.G., Cox, G.M. (1957). *Experimental Designs*, 2nd ed. John Wiley & Sons, New York.

Conover, W.J. (1980). *Practical Nonparametric Statistics*, 2nd ed. John Wiley & Sons, New York.

Cox, D.R. (1972). Regression models and life tables. *Journal of the Royal Statistical Society* 34:187–220.

Cronin, A.I., Bridge, C., Field, J. (2011). Climatic correlates in the temporal demographic variation in the tropical hover wasp *Liostenogaster flavolineata*. *Insecte Sociaux* 58:23–29.

Cryer, J.D. (1986). *Time Series Analysis*. Duxbury Publishers, Boston.

Davies, N.B., Krebs, J.R., West, S.A. (2012). *An Introduction to Behavioural Ecology*, 4th ed. Wiley-Blackwell, Chichester.

Desu, M.M., Raghavarao, D. (1990). *Sample Size Methodology*. Academic Press, Inc., Boston.

De Silva, S. (2010). Acoustic communication in the Asian elephant, Elephas maximus maximus. *Behaviour* 147:825–852.

De Silva, S., Ranjeewa, A.D.G., Kryazhimskiy, S. (2011). The dynamics of social networks among female Asian elephants. *BioMed Central Ecology* 11(17):1–15.

Draper, N.R., Smith, H. (1998). *Applied Regression Analysis*, 3rd ed. John Wiley & Sons, New York.

Dugatin, L.A., Reeve, H.K. (1998). *Game Theory and Animal Behavior*. Oxford University Press, New York.

Efron, B. (1982). *The Jackknife, the Bootstrap and Other Resampling Plans*. Society for Industrial and Applied Mathematics, Philadelphia.

Everitt, B.S., Landau, S., Leese, M., Stahl, D. (2011). *Cluster Analysis*. John Wiley & Sons, Chichester.

Fleiss, J.L., Levin, B., Paik, M.C. (2003). *Statistical Methods for Rates and Proportions*, 3rd ed. John Wiley & Sons, New York.

Fosler-Lussier, E. (1998). Markov models and hidden Markov models: A brief tutorial. TR-98-041, International Computer Science Institute, Berkley.

Gelman, A., Carlin, J.B., Stern, H.S., Rubin, D.B. (2000). *Bayesian Data Analysis*. Chapman & Hall/CRC, Boca Raton.

Golding, Y.C., Edmunds, M. (2000). Behavioural mimicry of honeybees (Apis mellifera) by droneflies (Diptera: Syrphidae: Eristalis spp.). *Proceedings of the Royal Society of London B* 267:903–909.

Good, P. (1994). *Permutation Tests*. Springer, Berlin.

Grimmett, G., Stirzaker, D. (2004). *Introduction to Probability and Random Processes*, 3rd ed. Oxford University Press, Oxford.

Hinde, C.A. (2006). Negotiation over offspring care? A positive response to partner-provisioning rate in great tits. *Behavioral Ecology* 17:6–12.

Hoeting, J.A., Madigan, D., Raferty, A.E., Volinsky, C.T. (1999). Bayesian model averaging: A tutorial. *Statistical Science* 14(4):382–417.

Hosmer, D.W., Lemeshow, S. (1989). *Applied Regression Analysis*. John Wiley & Sons, New York.

Hurlbert, S.H. (1984). Pseudoreplication and the design of ecological field experiments. *Ecological Monographs* 54(2):187–211.

Jeffreys, H. (1961). *Theory of Probability*. Oxford Press, London.

Johnson, N.L., Kotz, S., Kemp, A.W. (1992). *Univariate Discrete Distributions*, 2nd ed. John Wiley & Sons, New York.

Judge, G.G., Griffiths, W.E., Hill, R.C., Lutkepohl, H., Lee, T.C. (1985). *The Theory and Practice of Econometrics*, 2nd ed. John Wiley & Sons, New York.

Kalcenik, A. (1984). Central place foraging in starlings (Sturnus vulgaris). I. Patch residence time. *Journal of Animal Ecology* 53:283–299.

Kaplan, E.L., and Meier, P. (1958). Nonparametric estimation from incomplete observations. *Journal of the American Statistical Association* 53:457–481.

King, L. E., Soltis, J., Douglas-Hamilton, I., Savage, A., and Vollrath, F. (2010). Bee threat elicits alarm call in African elephants. *PLoS ONE* 5:e10346. doi: 10.1371 /journal.pone.0010346.

Law, A.M., Kelton, W.D. (2015). *Simulation Modeling and Analysis*, 5th ed. McGraw-Hill, New York.

Lee, E.T. (1992). *Statistical Methods for Survival Data Analysis*, 2nd ed. John Wiley & Sons, Inc., New York.

Mallows, C.L. (1973). Some comments on C_p. *Technometrics* 15:661–675.

Mann, N.R., Schafer, R.E., Singpurwalla, N.D. (1974). *Methods for Statistical Analysis of Reliability and Life Data*. John Wiley & Sons, New York.

Maynard Smith, J. (1982). *Evolution and the Theory of Games*. Cambridge University Press, Cambridge.

McQuaid, C.D. (1994). Feeding behavior and selection of bivalve prey by *Octopus vulgaris* Cuvier. *Journal of Experimental Marine Biology and Ecology* 177(1994): 187–202.

McQueen, J.B. (1967). Some methods for classification and analysis of multivariate observations. *Proceedings of the 5th Berkeley Symposium on Mathematical Statistics and Probability*, University of California, Berkeley, 1:281–297.

Montgomery, D.C. (2001). *Design and Analysis of Experiments*, 5th ed. John Wiley & Sons, New York.

Pardo, M.A., Pardo, S.A., Shields, W.M. (2014). Eastern gray squirrels (Sciurus carolinensis) communicate with the positions of their tails in an agonistic context. *The American Midland Naturalist* 172(2):359–365.

Pardo, S.A., Pardo, Y.A. (2016). *Empirical Modeling and Data Analysis for Engineers and Applied Scientists*. Springer, Basel.

Pearson, K. (1900). On the criterion that a given system of deviations from the probable in the case of a correlated system of variables is such that is can be reasonably supposed to have arisen from random sampling. *Philosophical Magazine Series 5* 50(302):157–175.

Peto, R., Peto, J. (1972). Asymptotically efficient rank invariant test procedures. *Journal of the Royal Statistical Society* 135(2):185–207.

Poole, J.H. (2011). Behavioral contexts of elephant acoustic communication. In *The Amboseli Elephants: A Long-Term Perspective on a Long-Lived Mammal*, pp. 125–159.

Poole, J.H., Payne, K., Langbauer, J., William, R., and Moss, C.J. (1988). The social contexts of some very low-frequency calls of African elephants. *Behavioral Ecology and Sociobiology* 22:385–392. doi: 10.1007/BF00294975.

Quinn, J., Tolson, K.M. (2009). Proximate methods of parasite egg rejection in Northern mockingbirds. *The Wilson Journal of Ornithology* 121(1):180–183.

Reklaitis, G.V., Ravindran, A., Ragsdell, K.M. (1983). *Engineering Optimization: Methods and Applications*. John Wiley & Sons, New York.

Romano, A., Bazzi, G., Caprioli, M., Corti, M., Costanzo, A., Rubolini, D., Saino, N. (2016). Nestling sex and plumage color predict food allocation by barn swallow parents. *Behavioral Ecology* 27(4):1198–1205.

Roth, M. (2013). On the multivariate t distribution. Technical Report LiTH-ISY-R-3059, Division of Automatic Control, Linkopings Univerisitet, Linkoping.

Rubin, D.B. (1976). Inference and missing data. *Biometrika* 63(3):581–590.

Rutherford, E. (1900). A radioactive substance emitted from thorium compounds. *Philosophical Magazine* 5:1–14

Saporito, R.A., Zuecher, R., Roberts, M., Gerow, K.G., Donnelly, M.A. (2007). Experimental evidence for aposematism in the dendrobatid poison frog *Oophagia pumilio*. *Copeia* 4:1006–1011.

Schürch, R., Heg, D. (2010). Life history and behavioral type in the highly social cichlid *Neolamprologus pulcher*. *Behavioral Ecology* 21(3):588–598.

Searle, S.R., Casella, G., McCulloch, C.E. (1992). *Variance Components*. John Wiley & Sons, New York.

Sinervo, B., Lively, C.M. (1996). The rock-paper-scissors game and the evolution of alternative male strategies. *Letters to Nature* 380:240–243.

Soltis, J., King, L.E., Douglas-Hamilton, I., Vollrath, F., and Savage, A. (2014). African elephant alarm calls distinguish between threats from humans and bees. *PLoS ONE* 9. doi: 10.1371/journal.pone.0089403.

Strang, G. (2016). *Linear Algebra and its Applications*, 5th ed. Wellsley-Cambridge Press, Cambridge.

Tadelis, S. (2013). *Game Theory: An Introduction*. Princeton University Press, Princeton.

Van Buuren, S. (2012). *Flexible Imputation of Missing Data*. CRC/Chapman & Hall, Boca Raton.

Wajnberg, E., Haccou, P. (2008). Statistical tools for analyzing data on behavioral ecology of insect parasitoids. In *Behavioral Ecology of Insect Parasitoids*, eds. E. Wajnberg, C. Bernstein, J. Van Alphen. Blackwell Publishing, Malden.

Webster, H., McNutt, J.W., McComb, K. (2011). African wild dogs as a fugitive species: Playback experiments investigate how wild dogs respond to their major competitors. *Ethology* 1–10.

Wedderburn, R.W.M. (1974). Quasi-likelihood functions, generalized linear models, and the Gauss-Newton method. *Biometrika* 61(3):439–447.

Whitehead, H. (2008). *Analyzing Animal Societies: Quantitative Methods for Vertebrate Social Analysis*. The University of Chicago Press, Chicago and London.

Whitten, K.W., Davis, R.E., Peck, M.L., Stanley, G.G. (2004). *General Chemistry*, 7th ed. Brooks/Cole, Belmont.

Wilson, E.O. (1980). Caste and division of labor in leaf-cutter ants (Hymentopera: Formicidae: *Atta*). *Behavioral Ecology and Sociobiology* 7:143–156.

Winkler, R. (1972). *Introduction to Bayesian Inference and Decision*. Holt, Rinehart, and Winston, Inc., New York.

Yekutieli, D., Benjamini, Y. (1999). Resampling-based false discovery rate controlling multiple test procedures for correlated test statistics. *Journal of Statisitcal Planning and Inference* 82(1–2):171–196.

Index

Lightning Source UK Ltd.
Milton Keynes UK
UKHW021136010223
416285UK00028B/373